PHOSPHORUS, FOOD, AND OUR FUTURE

Phosphorus, Food, and Our Future

Edited by Karl A. Wyant, Jessica R. Corman, James J. Elser

Oxford University Press is a department of the University of Oxford.
It furthers the University's objective of excellence in research, scholarship,
and education by publishing worldwide.

Oxford   New York
Auckland   Cape Town   Dar es Salaam   Hong Kong   Karachi
Kuala Lumpur   Madrid   Melbourne   Mexico City   Nairobi
New Delhi   Shanghai   Taipei   Toronto

With offices in
Argentina   Austria   Brazil   Chile   Czech Republic   France   Greece
Guatemala   Hungary   Italy   Japan   Poland   Portugal   Singapore
South Korea   Switzerland   Thailand   Turkey   Ukraine   Vietnam

Oxford is a registered trademark of Oxford University Press in the UK and certain other
countries.

Published in the United States of America by
Oxford University Press
198 Madison Avenue, New York, NY 10016

© Oxford University Press 2013

All rights reserved. No part of this publication may be reproduced, stored in a
retrieval system, or transmitted, in any form or by any means, without the prior
permission in writing of Oxford University Press, or as expressly permitted by law,
by license, or under terms agreed with the appropriate reproduction rights organization.
Inquiries concerning reproduction outside the scope of the above should be sent to the
Rights Department, Oxford University Press, at the address above.

You must not circulate this work in any other form
and you must impose this same condition on any acquirer.

Library of Congress Cataloging-in-Publication Data

Phosphorus, food, and our future / edited by Karl A. Wyant, Jessica R. Corman, James J. Elser.
pages cm
ISBN 978-0-19-991683-2 (alk. paper)
1. Phosphorus. 2. Phosphorus cycle (Biogeochemistry) 3. Food--Phosphorus content. I. Wyant, Karl A., editor of
compilation. II. Corman, Jessica R., editor of compilation. III. Elser, James J., editor of compilation.
QD181.P1P484 2013
333.8'56--dc23
                                    2013003194

1 3 5 7 9 8 6 4 2
Printed in the United States of America
on acid-free paper

Contents

Preface   vii
   Jessica R. Corman, Karl A. Wyant, Lisa Taylor, David Iwaniec, Rebecca Hale
Contributors   xiii
List of Abbreviations   xvii

Chapter 1: Introduction to P sustainability
P Is for Philosophy and Process   1
   Genevieve S. Metson, Karl A. Wyant, Daniel L. Childers

Chapter 2: The Biology and Ecology of Phosphorus in Biota, Natural Ecosystems, and Agroecosystems
P Is for Phosphorus   20
   James J. Elser, William M. Roberts, Philip M. Haygarth

Chapter 3: Global Phosphorus: Geological Sources and Demand-Driven Production
P Is for Price   40
   Donald Burt, Marion Dumas, Nathaniel Springer, David A. Vaccari

Chapter 4: Sustainable P in Agriculture: Food and Fuel
P Is for Productivity   64
   Val H. Smith, Cecil W. Forsberg, Roberto A. Gaxiola, Thomas W. Crawford, Jr., Andrew R. Sharpley, Laura Schreeg, Ben Chaffin

Chapter 5: Phosphorus in Urban and Agricultural Landscapes
P Is for Preservation   86
> Shelby H. Riskin, Gaston Small, Robert Mikkelsen, Genevieve Metson, Anna Bateman, James Cooper, Ola Stedje Hanserud, Philip M. Haygarth, Cecilia Laspoumaderes, Michelle McCrackin, Sonya Remington

Chapter 6: Phosphorus Recovery and Reuse
P Is for Processing   112
> Hiroko Yoshida, Kimo van Dijk, Aleksandra Drizo, Steven W. van Ginkel, Kazuyo Matsubae, Mark Buehrer

Chapter 7: Cultural Beliefs, Values, and the Biogeochemical Cycling of P
P Is for People   142
> Timothy Crews, James Cotner, Carol McCreary

Chapter 8: How MFA, Transdisciplinarity, Complex Adaptive Systems Thinking, and Education Reform Are Keys to Better Managing P
P Is for Parity   167
> Rebecca Cors, Kazuyo Matsubae, Anita Street

Chapter 9: Future Scenarios for the Sustainable Use of Global Phosphorus Resources
P Is for Preferred (p)Futures   183
> Daniel L. Childers, Zachary Caple, Cynthia Carlielle-Marquet, Dana Cordell, Vanda Gerhart, David Iwaniec, Stuart White

Chapter 10: Concluding Remarks: Synthesis and Initial Steps toward a Sustainable Phosphorus Future
P Is for Planning   199
> Jessica R. Corman, Karl A. Wyant, James J. Elser

Glossary   213
Index   221

Preface

Phosphorus is a unique chemical element, well-known for its biochemical role in binding DNA, its agricultural role as a fertilizer, and its environmental role as a pollutant. Over the past few years, we have begun to understand its complex economic and geopolitical role in our global society, as well. These insights have been aided by sustainability science, a perspective that incorporates society, economy, and environment into the management and decision-making process. The opportunity to push forward the heretofore nascent discussion inspired a group of ecology and sustainability science students to run a transdisciplinary conference combining sustainability themes with the science of phosphorus. In the spring of 2011, approximately 100 scientists, engineers, educators, policy makers, and professionals also heeded the call and came to Arizona State University (ASU) to discuss phosphorus sustainability.

NOTES ON THE CONFERENCE

Our goal for the conference was to facilitate cross-disciplinary and cross-sector discussions to define phosphorus sustainability issues and solutions from multiple stakeholder perspectives to create a discussion-based setting, filled with voices that represented the numerous stakeholders and disciplinary backgrounds involved with phosphorus sustainability. We achieved this by hosting a workshop-based conference. To maximize participant engagement, we solicited workshop ideas from the diverse participants as part of the registration process. A team of conference organizers then clustered and synthesized the suggested ideas into eight workshops. We assigned two facilitators to each workshop, usually a junior and senior participant together, from the participants who had indicated they would be willing to facilitate on their registration.

The list of workshops was then submitted to conference participants to select which workshop they would like to attend. Most participants were placed into their first workshop choice, and organizers ensured a mix of sectors (i.e., academia, NGO, private, and government) and disciplines were represented in each workshop. The eight workshops were: 1. Global distribution and demand of phosphate resources; 2. Role of agriculture in driving phosphorus demand; 3. Managing phosphoros in urban and agricultural landscapes; 4. Social implications of environmental services and disservices of phosphorus mining and use; 5. Strategies to improve the mitigation and recycling of the phosphorus waste stream; 6. Role that sustainable thinking across long time horizons could play in managing global phosphorus supplies; 7. The nexus between phosphorus supplies and food security; 8. Role that international policy will have in shaping sustainable phosphorus management.

The topical workshops ran concurrently and met for over 10 hours during the conference. In order to synthesize across topical domains and prevent workshop overlap, we also organized a mid-conference "workshop shuffle" during which participants formed new groups that had at least one member from each of the topical workshops. These groups met for one hour to report back on their workshops and get feedback from other groups.

Finally, at the close of the conference, all topical workshops reported back to the whole conference on their discussions, progress, and anticipated products (including, but not limited to, a consensus statement, which is presented on the next page, and chapters published in this volume). Of the eight workshops, seven produced chapters to contribute to this book.

Preface

frontiers in life sciences
Consensus Statement: Sustainable Phosphorus Summit

From 3-5 February 2011, more than 100 scientists, engineers, entrepreneurs, farmers, policy-makers, educators, artists, and others took part in The Sustainable Phosphorus Summit on the campus of Arizona State University in Tempe. The event was part of an emerging global dialogue around the diverse dimensions of human phosphorus use.

The Phoenix Phosphorus Declaration

We have achieved broad agreement on important issues surrounding phosphorus sustainability challenges and opportunities and seek to raise global awareness about them among all those with a stake in the future of food, water, and the biosphere. All of humanity, and indeed all living species, has this stake.

We find:
- **Essential and limited.** *Phosphorus is essential for all life because it is part of critical molecules like DNA. It is a limited natural resource needed to sustain the vitality and productivity of all ecosystems,* including farms.
- **Imbalanced cycle.** *Mining of phosphorus for fertilizer production has massively altered the cycling of phosphorus on Earth.* This increased phosphorus use has greatly expanded global capacity for food production but also has led to amplified phosphorus losses from cities, towns, and farms that can lead to degraded water quality, impair freshwater and marine fisheries, and alter natural biodiversity.
- **Food security.** *Phosphorus has a key role in global food security,* as reliable access to affordable fertilizer can allow farmers to improve yields and increase quality of life, especially in the developing world.
- **Recycle and reuse.** Currently, much phosphorus is lost in crop waste, food spoilage, and animal & human waste. *Recycling this phosphorus can reduce geopolitical and other uncertainties surrounding phosphorus fertilizer markets* and enhance farmer prosperity.
- **Reduce demand.** *Phosphorus natural resources can be extended by improving efficiency of use in agriculture, reducing erosion, limiting losses in mining & industry, and eating lower in the food chain.*
- **Interconnected.** *Phosphorus stewardship is coupled to other major global sustainability challenges,* including those involving energy, water, and other chemical elements.
- **Entrepreneurship.** There are great *economic opportunities to innovate and create new industries* for phosphorus supply diversification and for improved agricultural phosphorus efficiency. However, the suitability of such measures will differ for different environments, cultures, and contexts.

By closing the human phosphorus cycle and transforming wastes into resources and uncertainty into security, humanity can implement a "new alchemy" in which people become more secure and enjoy greater well-being in a healthy environment.

Participants in the 2011 Sustainable Phosphorus Summit Tempe, Arizona, USA

~ End of Declaration ~

For more information, visit:
sols.asu.edu/frontiers/2011/consensus.php

Email: sustainablePsummit@gmail.com

ASU School of Life Sciences is a division of the College of Liberal Arts and Sciences • Design by ASU School of Life Sciences Visualization Group

NOTES FROM THE ORGANIZING COMMITTEE

Lisa Taylor, Lead for the Art Outreach

In August of 2010, I knew virtually nothing about phosphorus sustainability. When I joined the Sustainable P Summit (SPS) organizing committee, my main role was to lead an outreach event that would engage the public and extend the impact of the Summit beyond the scientific community. One of my first responses to telling a friend about my new job: "Really? An outreach event about phosphorus? That doesn't sound very fun." But the SPS team that had conceived of the idea for the conference and already secured funding was an enthusiastic, hardworking, and innovative bunch of PhD students with important ideas to share. If we could combine their science and their passion with a creative way to communicate it, we could teach the public about phosphorus. And everyone might even have fun.

We decided on a collaborative art exhibit with a simple goal: to pair artists and scientists to communicate ideas about phosphorus sustainability to the public. We sent out a call to artists and the responses were overwhelmingly positive. Artists were excited about the opportunity to meet and work directly with an expert on phosphorus sustainability. Scientist collaborators from around the world (recruited during the P Summit registration process) were equally excited about the chance to work with an artist to communicate their research. We created twenty artist–scientist teams to create work for the exhibit. The final show featured multimedia and interactive installations, paintings, monoprints, collage, cut paper, illustrations, photographs, fashion, dance, music, and even a decoupage toilet—all with an important message to share about P.

The show was exhibited at the Desert Botanical Garden in Phoenix, Arizona, during the final evening of the Sustainable P Summit, where it was viewed by more than 300 guests from the scientific community and the public. The work then traveled to the Step Gallery at ASU's Herberger Institute for Design and the Arts in Tempe, Arizona, where it was again showcased free for the public. I invite you to check out our online gallery where we will display this work indefinitely. In addition, the authors of each chapter of this book have selected a piece from the exhibit that represents the themes they cover in the text. We hope that having artwork interspersed throughout the book will provide frequent inspiration to scientists to keep artists involved in the conversation.

KARL A. WYANT, HEAD EDITOR FOR THE BOOK PROJECT

When the question "Who wants to develop the results of our workshop collaborations into a book?" came up in the spring of 2010, the meeting room went silent as folks mulled over their decision. Nobody in the room had ever written a book proposal, shopped a book for publication, or served as head editor on such a large project before. This was new territory for the members of the graduate student planning committee. After a few seconds of silence, I spoke up and agreed to be the lead on this project. The rest of the room sighed in relief as somebody (i.e., not them) decided to throw their hat in the ring for an adventure into the unknown.

Looking back on it now, I don't really know why I decided to sign up as the head editor of the book. Perhaps, I had been in the lab too long and my brain was a bit fuzzy and confused. Honestly, the real reason I signed up was that it sounded like a challenging and rewarding experience and, potentially, a lot of fun. I spent the summer and fall of 2010 writing book proposals, cold calling and e-mailing book publishers, and blindly fumbling through the mysterious new world of heading up a book project. Meanwhile, we had a huge international conference coming up in the spring of 2011 that needed planning. Thankfully, I was involved with a hardworking, enthusiastic group of my peers. As our workshop ideas coalesced and solidified, so did the emerging chapter themes of the book. I remember creating a brief outline of the book structure, and with a giant leap, I then made an author style guide and started soliciting conference participants for their involvement with authoring the individual book chapters. Secretly, I was worried that nobody would want to come on board with this ambitious project with such a greenhorn at the helm. With a nervous click of the mouse, I sent out the author style guide and gingerly waited for my turn to make "book announcements" on the second day of the Sustainable P Summit.

I was pleasantly surprised with the enthusiasm expressed by the book collaborators. Folks were excited to work on this book, As the conference wore on, our book, no longer a pipe dream living on my hard drive, was now being shaped and kneaded by the international phosphorus community. By the end of the conference, chapter ideas were proposed and solidified, the book theme was strengthened, and most importantly, we set a series of deadlines for the completion of book drafts.

Enter Dr. James Elser and Ms. Jessica Corman. After our book idea gained traction and began to mature at the P Summit, I soon realized I needed some help managing and keeping the project moving forward. Dr. Elser had previously published a textbook and quickly set about using his experience to secure a publisher, Oxford University Press. Jessica Corman, who was the head of the P Summit planning committee, was a wizard at managing large projects. Things were looking great for the book that had only existed as an idea on paper months earlier. We had our authors on board, we had the publisher, and we had our editing team; now we just needed to keep our momentum going.

The book you see now really hit its stride when we had a second meeting in late August in Arizona. I know what you are thinking: "Arizona! In August? You are crazy!" Our goal was to invite all the authors back for an intense, three-day writing session that would allow authors to work one-on-one with each other and also allow chapter working groups to read through the nearly completed drafts of every chapter. Our goal was to streamline and smooth out the themes of the book in person rather than sending out the endless barrage of e-mails one might expect from a project like this. How, you might ask, can we expect our authors to work so intensely in the brutal Arizona heat of late August? The editing team had one more trick up its sleeve.

Our second reunion, or the Echo Meeting, as we called it, was held in the cool mountain pine forests of Arizona's Mogollon rim country. Here, in the refreshing summer rains, we worked into the late hours of the night refining, revising, and in some cases, totally reworking the chapters that you see in the book today. Our book, which had existed just as a rough outline and author style guide six months prior, was now maturing before our eyes.

As you read through the chapters of the book, I invite you to keep in mind that the origins of this project came from the minds of graduate students sitting around a table at Arizona State University nearly two years ago. Graduate students conceived a conference idea that we felt strongly about and, with great zeal, began to shape the events of the Sustainable P Summit and, ultimately, the book that you are holding in your hands. If there is any lesson in all of this, it is that little ideas can become big ideas given the right amount of hard work, diligence, and the involvement of enthusiastic, knowledgeable team members.

Contributors

Anna Bateman
Department of Civil Engineering, University of Birmingham, Edgbaston, Birmingham, UK

Mark Buehrer
2020 Engineering LLC, Bellingham, WA

Donald Burt
School of Earth and Space Exploration, Arizona State University, Tempe, AZ

Zachary Caple
Department of Anthropology, University of California Santa Cruz, Santa Cruz, CA

Cynthia Carlielle-Marquet
School of Civil Engineering, College of Engineering and Physical Sciences, University of Birmingham, Edgbaston, UK

Ben Chaffin
Chaffin Farms, Ithaca, MI

Daniel L. Childers
School of Sustainability, Arizona State University, Tempe, AZ

James Cooper
Department of Civil Engineering, University of Birmingham, Edgbaston, Birmingham, UK

James Cotner
Department of Ecology, Evolution and Behavior, University of Minnesota— Twin Cities, St. Paul, MN

Dana Cordell
Institute for Sustainable Futures, University of Technology Sydney, Broadway, Australia

Jessica R. Corman
School of Life Sciences, Arizona State University, Tempe, AZ

Rebecca Cors
ETH Zürich, Institute for Environmental Decisions, Natural and Social Science Interface, Zürich

Thomas W. Crawford, Jr.
Bio Huma Netics, Inc., 1331 W. Houston Ave, Gilbert, AZ

Timothy Crews
The Land Institute, Salina, KS

Aleksandra Drizo
Plant and Soil Science Department, University of Vermont, Burlington, VT

Marion Dumas
Sustainable Development Doctoral Program, Columbia University, New York, NY

James J. Elser
School of Life Sciences, Arizona State University, Tempe, AZ

Cecil W. Forsberg
Department of Molecular and Cellular Biology, University of Guelph, Guelph, Canada

Roberto A. Gaxiola
Department of Cellular and Molecular Biosciences, Arizona State University, Tempe, AZ

Vanda Gerhart
AgroEcology and Food Systems, Tucson, AZ

Rebecca Hale
School of Life Sciences, Arizona State University, Tempe, AZ

Ola Stedje Hanserud
Norwegian Institute for Agricultural and Environmental Research (Bioforsk), Norway

Philip M. Haygarth
The Lancaster Environment Center, Lancaster University, Lancaster, England, UK

David Iwaniec
School of Sustainability, Arizona State University, Tempe AZ

Cecilia Laspoumaderes

Centro Cientifico Tecnológico Comahue del CONICET, Bariloche, Rio Negro, Argentina

Kazuyo Matsubae

Department of Metallurgy, Graduate School of Engineering, Tohoku University, Japan

Michelle McCrackin

Washington State University—Vancouver, Vancouver, WA; National Research Council, Research Associateship Program, Washington DC

Carol McCreary

Public Hygiene Lets Us Stay Human (PHLUSH), Portland, OR

Genevieve S. Metson

Department of Natural Resource Science, McGill University, Ste. Anne de Bellevue, QC, Canada

Robert Mikkelsen

International Plant Nutrition Institute, Merced, CA

Sonya Remington

School of Sustainability, Arizona State University, Tempe, AZ

William M. Roberts

The Lancaster Environment Center, Lancaster University, Lancaster, LA1 4YQ, England, United Kingdom & The James Hutton Institute, Craigiebuckler, Aberdeen AB15 8QH, Scotland, United Kingdom

Shelby H. Riskn

Department of Ecology and Evolutionary Biology, Brown University, Providence, RI; Ecosystems Center, Marine Biological Laboratory, Woods Hole, MA

Laura Schreeg

Department of Botany, University of Florida, Gainesville, FL

Andrew R. Sharpley

Department of Crop, Soil, and Environmental Sciences, University of Arkansas, Fayetteville, AR

Gaston Small

Department of Ecology, Evolution, and Behavior, University of Minnesota, St. Paul, MN

Val H. Smith

Department of Ecology and Evolutionary Biology, University of Kansas, Lawrence, KS

Nathaniel Springer

Dept. of Economics, Rensselaer Polytechnic Institute
Troy, NY 12180

Anita Street
United States Department of Energy, Washington, DC

Lisa Taylor
School of Life Sciences, Arizona State University, Tempe, AZ

David A. Vaccari
Dept. of Civil, Environmental and Ocean Engineering, Stevens Institute of Technology, Hoboken, NJ

Kimo van Dijk
Department of Soil Quality, Wageningen University & Research Center, Wageningen, The Netherlands

Steven W. van Ginkel
School of Civil and Environmental Engineering, Georgia Institute of Technology, Atlanta, GA

Stuart White
Institute for Sustainable Futures, University of Technology Sydney, Broadway, Australia

Karl A. Wyant
School of Life Sciences, Arizona State University, Tempe, AZ

Hiroko Yoshida
Department of Environmental Engineering, Denmark Technical University, Lyngby, Denmark

List of Abbreviations

CHAPTER 1

P	phosphorus

CHAPTER 2

ATP	adenosine tri-phosphate
DNA	deoxy-ribonucleic acid
N	nitrogen
P	phosphorus
PO_4	phosphate
RNA	ribonucleic acid

CHAPTER 3

IFA	International Fertilizer Industry Association
IFDC	International Fertilizer Development Center
DAP	Diammonium phosphate (fertilizer)
MAP	Monammonium phosphate (fertilizer)
OCP	Office Chérifien des Phosphates (Morocco)
PR	Phosphate rock
PR/N	Per capita annual phosphate rock production
R/C	Ratio of reserves to annual consumption rate
TSP	Triple super phosphate (fertilizer)
USGS	United States Geological Survey

CHAPTER 5

IPNI	International Plant Nutrition Institute
N	Nitrogen
K	Potassium
WWTP	Wastewater treatment plant

CHAPTER 6

BMP	Best management practice
EBPR	Enhanced biological phosphorus removal
FAO	Food and Agriculture Organization
MAP	Magnesium Ammonium Phosphate
OECD	Organisation for Economic Co-operation and Development
OFMSW	Organic Fraction of Municipal Solid Waste
OWTP	On-site wastewater treatment systems
P	Phosphorus
WWTP	Wastewater treatment process
WHO	World Health Organization
UNICEF	United Nations Children Fund
JMP	Joint monitoring program

CHAPTER 8

MFA	Material flow analysis
Global TraPs	Global Transdisciplinary Processes for Sustainable Phosphorus Management

PHOSPHORUS, FOOD, AND OUR FUTURE

1

Introduction to P Sustainability

P IS FOR PHILOSOPHY AND PROCESS

Genevieve S. Metson, Karl A. Wyant, Daniel L. Childers

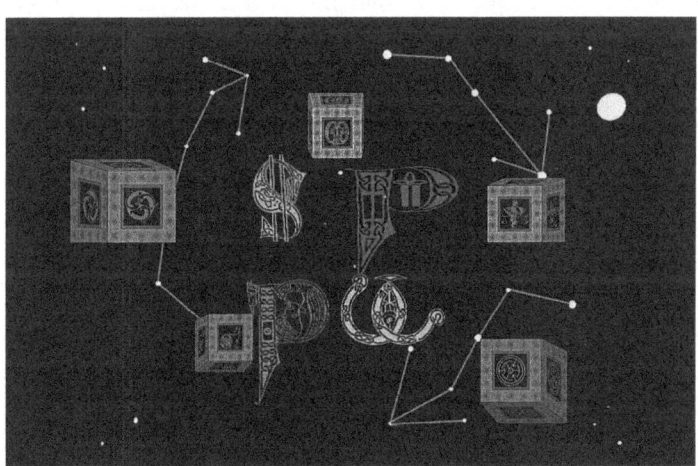

Blake McConnell; *The PTown Constellations;* Digital media various sizes

Blake McConnell
The PTown Constellations

Scientific Collaborator:
David Iwaniec, PhD student, School of Sustainability, Arizona State University

Description of Artwork:

The Ptown Constellations create a dynamic portrait of a theoretical world, like our own, facing a shortage of usable phosphorus. This compound, essential to agriculture, food production, and the fabrication of many materials, lies at the center of a growing debate about sustainable practices as they relate to economic and political priorities. Inspired by the PlanPHX visioning process, these constellations, while introducing the user to different types of sustainability strategies at different parts along the human phosphorus cycle, allows them to rank phosphorus conservation methods and see how resulting consequences compare to each other. Through interaction, they learn about phosphorus conservation issues and how personal, political, and economic choices affect others.

Users rank conservation methods using a physical game board, choosing from five choices represented in an LCD projection onto a table—investment in mining technology, phosphorus taxation, reduction of demand through consumer choices, genetic engineering, and closing the loop through recycling. Icons in a second display—representing increased cost, environmental pollution, political fallout, and even unknown consequences—portray the intensity of potential consequences to the user's ranking by growing or receding from view. This feedback alerts the user to consequences beyond phosphorus scarcity, allowing them to experiment with different strategies and resulting outcomes in pursuance of a normative state.

Each conservation method and consequence is represented by an icon. These icons, "sampled" from ancient Celtic artifacts and "hacked" using contemporary photo doctoring techniques, incorporate a design aesthetic that emphasizes interconnectedness. Mirroring the complexity of the human/phosphorus cycle, Celtic knots connect in unexpected ways, requiring close inspection to perceive their true path. These images, which reference human and animal forms, also recall patterns attributed to clusters of stars. Whereas our ancestors saw figures emerge from points in the night sky, we see them arise from our complex resource extraction and allocation systems. Do these figures appear as angels or monsters? The PTown Constellations let the viewer decide.

About the Artist:

Blake McConnell is a media artist, musician, and activist. Originally from the Atlanta suburb of Marietta, Georgia, and a longtime resident of San Francisco, California, he now resides in Phoenix, Arizona. His creative work manifests in various ways, though the intersection of media, technology, and society serves as its fulcrum. The confluence of the implications of form and the infinite variation of perception fuel his inquiry. Inspired by years of local organizing, he embraces the frontiers of new media while respecting their incumbent responsibilities.

BOX 1.1
CHAPTER 1 OBJECTIVES

- Illustrate and explain the basic principles of sustainability.
- Apply a sustainability framework to existing issues in P management.
- Summarize human impacts on natural P cycling and the identification of P sustainability as a "wicked problem."
- Describe how a sustainability perspective contributes to shaping appropriate solutions to P management.
- Describe the general approach of upcoming book chapters in connecting various aspects of human use of P to a sustainability framework.
- Apply a sustainability framework to existing issues in P management.
- Summarize human impacts on natural P cycling and the identification of P sustainability as a "wicked problem."
- Describe how a sustainability perspective contributes to shaping appropriate solutions to P management.
- Describe the general approach of upcoming book chapters in connecting various aspects of human use of P to a sustainability framework.

INTRODUCTION

The dynamics of natural resources, whether water, oil, or nutrients, can be difficult to explain, as well as manage. Natural cycling is layered with social and economic dynamics and the result can be quite complex, and exhibit a lot of spatial and temporal heterogeneity. Phosphorus (P) is no exception. Although complex, it is clear that current dynamics and P-use patterns are problematic. In fact we rarely directly manage P resources; instead we manage fertilizer, crop production, and water, which in turn affect P cycling. In this chapter, and throughout this book, we will bridge our current segmented way of viewing anthropogenic P cycling (e.g., mining, fertilizer, crop production, food consumption, and waste management) and systems-thinking view of sustainability. In this way, we hope to move toward solving challenges of current unsustainable P-use. In this chapter we will explain 1) what sustainability science is, 2) why the current state of P cycling is a sustainability challenge, 3) how we got to the current state of P cycling, and 4) how sustainability science frames the type of solutions we should consider when trying to change P cycling for the better.

A PRIMER ON SUSTAINABILITY SCIENCE

Many scientists and policy makers have recognized that traditional discipline-focused views of the world and of solving problems are no longer adequate for either

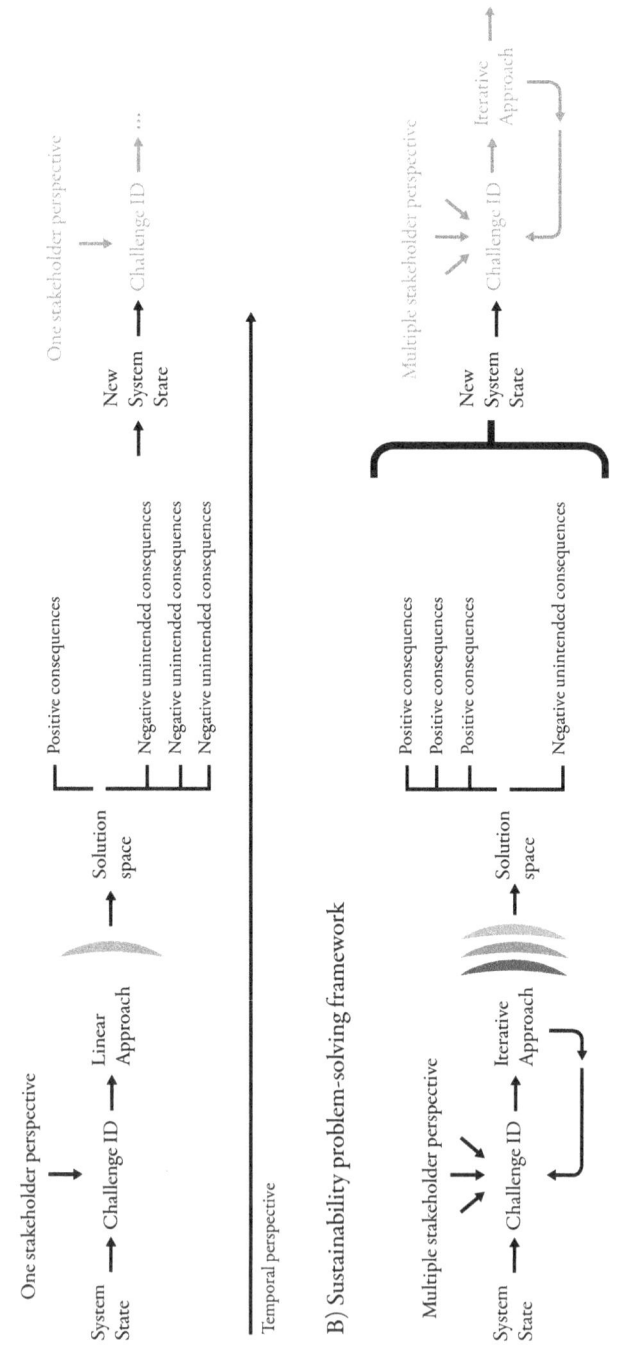

understanding or deriving solutions to complex problems (Figure 1.1, Funtowicz and Ravetz 1993). In many ways, **sustainability*** science is a response to increasing awareness about negative **externalities*** arising from the management of resources (and services) for a single goal, often developed by a single discipline. For example, clear-cutting forests may be the most efficient means to exploit that resource for timber production. However, there are unintended consequences to such a practice that affect society immediately, such as landslides and water pollution, as well as those that affect future generations, such as the eventual depletion of forest resources for future use. Indeed, the classic "Tragedy of the Commons" (Harding 1968) demonstrates pitfalls of linear thinking, where a focus only on narrow, often individual benefits, comes at the cost of the group.

The idea of "sustainability" began to formally emerge in the early 1970s with the publication of the Stockholm Declaration on the Human Environment, which led to the creation of the United Nations Environment Programme (UNEP, Declaration of the United Nations Conference on the Human Environment 1972). The UNEP sought to reconcile the desire for economic development and maintaining environmental protection (Dresner 2002). Fast-forward to the late 1970s, which brought the concept of a "sustainable society" founded on equity and participation in the democratic process. These ideas formed the cornerstone of "Our Common Future" (Brundtland 1987), also known as the Brundtland Report, which formally defined the term sustainability. This influential report set the stage for the modern field of sustainability science. Further developments throughout the 1990s included the Earth Summit I (1992) and II (1997) and the Kyoto Protocol (Dresner 2008). According to Agyeman et al. (2003), sustainability considers "the need to

---

FIGURE 1.1 Contrast between approaches and solutions when using a conventional problem-solving framework (**A**) vs. a sustainability problem-solving framework (**B**). (**A**) The system state is the current way a system works (which could be bounded by place, by sector, or by theme). If a challenge (or problem) is identified by one type of stakeholder and, when we use a linear approach to solving this challenge we will come up with one solution to that problem. This solution, however, only addresses instead a specific problem and may have missed the complex interactions that system state exhibited. This solution may thus only work for a short amount of time, and also result in a multitude of negative consequences. The link between the unintended consequences and the intervention may be missed, resulting in the same linear method to be applied to "new" challenges. (**B**) On the other hand, if multiple stakeholders participate in characterizing the system state and identify challenges their diverse perspectives can result in a more complete understanding of the challenge. This "sustainability lens" is really the multiple perspectives of stakeholders coming together. This framework will result in solutions that may be more appropriate to minimize negative consequences and maximize the number of benefits from the changes. In this way, the solutions are more long-term (hence the longer time arrow). Still, there may be unintended negative consequences. However, the system's perspective will ensure that linkages are understood and new solutions are found through an iterative process. Based on ideas and methods further developed in Scholz et al. (2006) and Robert et al. (2002).

ensure a better quality of life for all, now and into the future, in a just and equitable manner, whilst living within the limits of supporting ecosystems." The definition of, and methods applied in, sustainability are continually evolving. However, we believe Agyeman's definition represents the overarching themes of both sustainability science and sustainable development.

In order to understand complex social-ecological problems, and embrace sustainable development principles in our solutions, we must include and integrate the environmental, social, and economic components of a problem or challenge. A simple visualization of this is with three concentric spheres (Figure 1.2). The environment sphere is the all-encompassing life-support system upon which humans (and all other species) depend and is the largest of the three spheres. That said, sustainability is ultimately an anthropocentric concept where we aim to equitably support current and future human populations. Thus, society is an essential consideration in sustainability, including societal interactions with the larger environment and among people. Thus, the society sphere is within the environment sphere. The economy is a human construct, the tool by which societies manage many interactions (between people and nature, and among people), and is thus a sub-sphere of the societal domain. Because sustainability problems are complex, focusing on less than all three of these spheres is inadequate and inevitably produces simplistic and ineffective solutions (Figure 1.1a).

Sustainability researchers strive to understand social (sometimes social-economic) and environmental interactions, particularly the key feedbacks in **social-ecological systems*** (Kates et al. 2001, Sarewitz et al. 2010). Only by understanding these feedback dynamics can we intervene effectively to mitigate harmful effects and foster beneficial outcomes. Sustainability scientists also aim to participate in decision-making processes, as opposed to simply providing information to decision makers. Gibson (2006) proposed eight criteria to assess the sustainability of a system (Table 1.1), and these criteria can be used to identify the kinds of problems that sustainability research and practice are best suited to tackle. Problems that fit these criteria are complex, urgent, exhibit long-term dynamics, involve cross-sectoral and cross-scalar interactions, and often have solutions that are place-based; sustainability researchers refer to these as "**wicked problems***" (Conklin 2006).

FRAMING P-USE AS A SUSTAINABILITY CHALLENGE

In this book we suggest that P-use is one of the wicked problems for which a sustainability science "lens" is appropriate. This is because P cycling is involved in the provision of basic human needs such as sufficient food and clean water. The environmental stewardship necessary to satisfy these needs in the long term necessitates a

TABLE 1.1

SUSTAINABILITY ASSESSMENT CRITERIA MODIFIED FROM GIBSON (2006) THAT IDENTIFY THE TYPES OF PROBLEMS SUSTAINABILITY IS BEST SUITED TO TACKLE (I.E., SYSTEMS WHERE MANY OF THESE CRITERIA ARE NOT MET OR PROBLEMS THAT AFFECT THESE CRITERIA FOR GLOBAL AND LOCAL SOCIETIES) AND THE CHARACTERISTICS OF SOLUTIONS THAT SHOULD BE CONSIDERED (I.E., SOLUTIONS THAT SUPPORT ENVIRONMENT, SOCIETY, AND ECONOMY TOGETHER).

Sustainability Criterion	Definition	How Criterion Applies to P-Use and Related Chapter in This Book
Socio-ecological integrity	Supports socio-ecological interactions that conserve irreplaceable life-support systems	Eutrophication caused by P-use is threatening water quality (ch. 2 and ch. 5).
Livelihood sufficiency and opportunity	Ensures universal access to resources (natural and societal) needed to live a healthy and fulfilling life	Current access to P as fertilizer is unequal around the globe and farmers in poor countries do not have access to enough fertilizer to grow enough food to feed local populations and have a livelihood (ch. 3, ch. 4, and ch. 8).
Intragenerational equity	Supports equity in the capacity and opportunities regardless of economic status	There are large gaps between access to food and sanitation infrastructure between rich and poor communities. These gaps dramatically alter how specific places contribute to P requirements and pollution (ch. 7).
Intergenerational equity	Ensures equity in the capacity and opportunities between current and future generations	Our currently intensive use of mineral P and lack of alternative recycling infrastructure makes future generations extremely vulnerable to P-price fluctuations and scarcity (ch. 9).

(Continued)

TABLE 1.1 (*Continued*)

Sustainability Criterion	Definition	How Criterion Applies to P-Use and Related Chapter in This Book
Resource maintenance and efficiency	Maintains long-term integrity of life support systems (the biosphere)	Current use of P degrades coastal and lake environments, which are essential to regional ecosystem function (ch. 5).
Socio-ecological civility and democratic governance	Creates governance systems that support collective and responsive decision making	There is a lack of collective governance both between steps in human P-use (i.e., mining to food to waste) and across the globe to manage P resources (ch. 6).
Precaution and adaptation	Avoids decisions that may have irreversible consequences while embracing uncertainty with nimble and adaptive processes	Current mining of P and ultimate dispersal of the resource in ocean waters depletes mineral resources and prohibits P recovery downstream (ch. 3 and ch. 6).
Immediate and long-term integration	Supports interventions that mutually benefit today's and future needs and desires	Today's intensive use of non-renewable P resources puts the current benefit of the few over the benefit of food security and clean environment for future generations (ch. 9).

judicial and systematic management of P resources. In other words, P-use entails all eight of Gibson's sustainability criteria (2006). Current use of P as a non-renewable resource and current price structures prevent effective responses to P scarcity (see chapters 3 and 8) and P pollution (see chapters 2 and 4). Subsidization of P fertilizers (chapter 8), unequal distribution of P demand and P availability (chapters 2, 4, and 8), the essential role of P in food security (chapter 8), and the economic externalization of P pollution effects on aquatic ecosystems (chapter 2) all point to the clear conclusion that current approaches to P resources and P-use is inadequate, even failed. Unless we radically change how humanity impinges on global cycling of P, the problems will only worsen. In this section we will divide the P-use situation into selected environmental, social, and economic components to illustrate various connections to sustainability theory (we have also italicized key concepts brought up in both Gibson's sustainability criteria and the conceptualization of a "wicked problem").

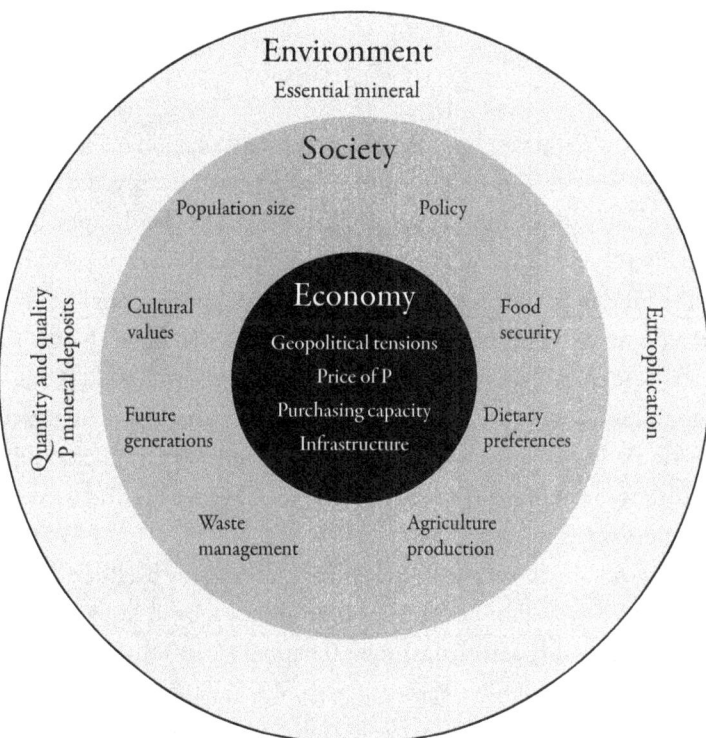

FIGURE 1.2 Phosphorus dynamics in relation to sustainability thinking. The image illustrates the three large considerations of sustainability: environment (light gray sphere), society (dark gray sphere), and economy (black sphere) as a nested model where environment encompasses society, and economy as part of society. Words in each sphere are elements that contribute to the phosphorus use problem.

Environment

There are no substitutes for P; it is biologically essential and is thus a necessary component of the *life-support system* that is our biosphere (chapter 2). Phosphorus is essential to human society (food and the production of other goods) as well as to all other species and the ecosystems that provide services to humans. Geological processes cycle P on *long time-scales* and this has led to the unequal distribution of concentrated mineral P deposits around the globe (chapter 3). Plant-available P is also unequally distributed across the globe because of different soil characteristics (chapters 2, 4, and 5, Cordell et al. 2009). The environment is also the recipient of the downstream consequences of inefficient and wasteful human uses of P (chapter 2). Impacts include the eutrophication of fresh and coastal water bodies (algal blooms that severely degrade water quality), as well as mining pollution (chapters 2, 3, and 5). These negative consequences have direct effects on people and economies by impairing various ecosystem services, including provision of safe drinking water, food provisioning through fisheries, as well as recreational amenities and habitat quality.

Society

Cultural norms and local biophysical characteristics of a region make both the problems (chapter 5) and the solutions (chapters 6 and 7) to P-use specific to a *particular time and place*. How different societies produce food, alter landscapes, choose their diet (especially vegetarian vs. meat intensive), and manage their wastes shapes their effects on the **human P cycle*** (per Childers et al. 2011). Interestingly, only about 20 percent of all P used to produce food is actually consumed in food—the remaining 80 percent is lost to inefficiencies and waste in the human P cycle (Childers et al. 2011). Beyond farm losses of P (which are explored in chapter 5), about 55 percent of the P in food is lost to inefficiencies "between farm and fork," including wastes in processing, transportation, and storage (waste management is discussed in chapter 6, Cordell et al. 2009).

Different people contribute to, and are affected differently by, P-use and P cycling. For example, the role of P in limiting crop production or in impairing water supplies differs considerably across the globe (chapter 8). In sub-Saharan Africa and in most countries with highly weathered soils, P is the limiting nutrient for plant growth. Agricultural production in these regions requires increases in P fertilizer application to maintain high yields for increasing human populations (Drechsel et al. 2004, Van Wambeke et al. 2004). At the same time, in many parts of the United States and Europe, P has been over-applied for many years, leading to high concentrations of P in agricultural soils and runoff, and consequent freshwater eutrophication (Bennett et al. 2001; Carpenter et al. 1998, see chapters 4 and 5 for more discussion). While the benefits of changing patterns of P-use to reduce downstream pollution may be more important to more developed countries, the effect of P-use on the price and availability of P may be more important to less developed countries.

Economy

Globally, most economically viable mineral P resources are controlled by only a few countries. This makes current and future P availability and accessibility very uncertain. Geopolitical tensions may affect the availability of P resources around the world and these tensions will only increase as supply declines and demand increases, causing prices to increase—perhaps dramatically (chapter 3). The problem of P supply is thus less one of geologic scarcity than of economic or geopolitical scarcity. Regions vary both in access to P and access to capital, labor, and technology to deal with P scarcity and eutrophication (chapter 8). For example, the current fraction of total farm operating costs that fertilizer purchases represent dramatically alters the effect of P price fluctuations on food systems and the response farmers and consumers have to such price fluctuations. Compared with Europe, P fertilizer is more expensive in sub-Saharan Africa—both in real price and as a proportion of a farm's budget—where sub-Saharan farmers have relatively less purchasing power. This means that **P accessibility*** for a sub-Saharan African farmer is considerably lower than for a European farmer, even though both are using mineral P from the same source (Cordell et al. 2009). Because of this disparity, traditional market forces (e.g., fertilizer prices) may not begin to enforce P conservation in richer countries until long after there is considerable food scarcity in the developing world. The lessons of the 2008 food crisis demonstrate that political instability often accompanies such adversity.

In summary, global P dynamics are a result of complex interactions of environmental, social, and economic factors operating at many scales. The diversity of local needs and the capacity to deal with P scarcity or P pollution challenges highlight the need for a sustainability approach that does not use "one-size-fits-all" solutions. However, before looking to the future, it is important to understand how the problematic P cycling we have identified came to be. Such a historical perspective is essential to understand what has influenced us to utilize P in the way we have and subsequently identify key aspects of resource, farm, and food management that need to be altered in moving forward (chapter 9).

HOW WE GOT TO THE CURRENT STATE OF P-USE

Early farming relied on P already in soils, on natural P inputs (such as sedimentation associated with flooding and chemical weathering of parent material), and on tight and efficient recycling of P (Ashley et al. 2011). In this early human P-cycle, P cycled much more conservatively relative to modern agro-ecosystems. Low-impact (i.e., less irreversible or large-scale changes in ecosystems in and around agricultural areas) food production was largely sustainable for thousands of years, with little need for outside nutrient additions, and had limited long-term downstream effects (Filippelli et al. 2008). The levels of P in any soil are highly localized and dependent on climate, topography, rates of **pedogenesis***, ecosystem age, and bedrock characteristics. As such, early societies developed methods of agriculture that worked with their local climates and soil types. However, such systems were far from perfect, as famines were more frequent and agriculture was mostly subsistent or limited in economic scope (Ashley et al. 2011). As P levels declined in soils (via leaching, occlusion, or crop uptake), communities dependent on these soils would either employ measures such as burning their fields to unlock bound P (Cordell 2001), wait for spring floods to renew their soil (e.g., annual silt deposits along the Nile River), or, in some cases, farmers were forced to find more fertile fields in other locations.

As human populations grew (and became more sedentary), soil amendments became necessary to maintain soil fertility (Ashley et al. 2011). The use of human and animal excreta (particularly from herbivores) on fields has long been practiced, particularly in China (Liu et al. 2008). Middle Eastern desert dwellers maintained collections of pigeons not only for their meat but, perhaps more importantly, for their P-rich guano, which was applied to soils that supported fruit trees and small gardens (Tepper 2007). These are just a few examples of locally sourced applications of P amendments in early agriculture. As reflected throughout this book, these systems of P application gave way in modern times to linear rather than circular modes of crop fertilization, food consumption, and waste management due to increasing demands for inexpensive food.

Starting in roughly the nineteenth century, global use of P switched from an **open cycle*** to a **closed cycle.*** Localized P sources were now sourced from locales that were much more distant from their intended destination (Figure 1.3), especially in Westernized agriculture. With urbanization, food consumption and waste production increasingly took place at some distance from crop production. Excreta were no longer systematically returned to farm fields due to the advent of modern sanitation practices (chapter 7). Furthermore, animal and feed production also became separated, as agricultural systems industrialized and farms specialized, leading to reduced on-field manure recycling. Countries such as England and the United States, in

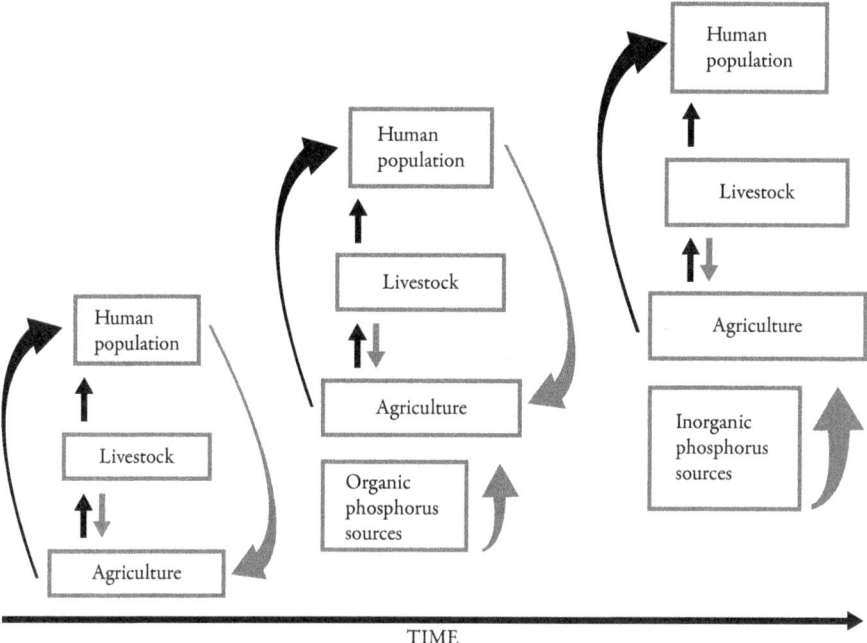

FIGURE 1.3 Changes in phosphorus inputs from 1800 to 2010. Black arrows represent flows of P from one pool to another. Gray arrows represent a return of P to a previous pool. The source of P inputs to agricultural fields has changed considerably from a largely localized, organic source (e.g., night soil, animal manures, crop residues) to a source largely dominated by inorganic sources such as mineral P. Figure created from concepts put forth in Ashley et al. 2011.

order to create organic liquid fertilizers, imported P as animal bones treated with sulfuric acid (Ashley et al. 2011). In the mid-1800s, bat and sea bird guano was mined and imported from South America and islands in the South Pacific (Ashley et al. 2011). As these sources were quickly exhausted and became increasingly expensive to process and transport, mineral P from mining soon became the fertilizer of choice for modern agriculture, due to its higher concentration of P available for crop use relative to organically sourced P like manures (Figure 1.4). Agricultural demand during the last 75 years—a result of the **Green Revolution***—increased global P mobilization by roughly fourfold (Filippelli 2008). During this time human population doubled and the consumption of meat and dairy foods increased (Falkowski et al. 2000, Villalba et al. 2008). Through this revolution, the nature of agricultural operations changed toward large, commercial monoculture production (which we refer to as the "agro-industrial complex") in response to increasing demand for food, fiber, and meat products (although small-scale farming still persists, especially in developing countries). The separation of P supply and demand, which characterized the development of this agro-industrial complex, was largely made possible by the use of cheap fossil fuels that characterized the twentieth century, making

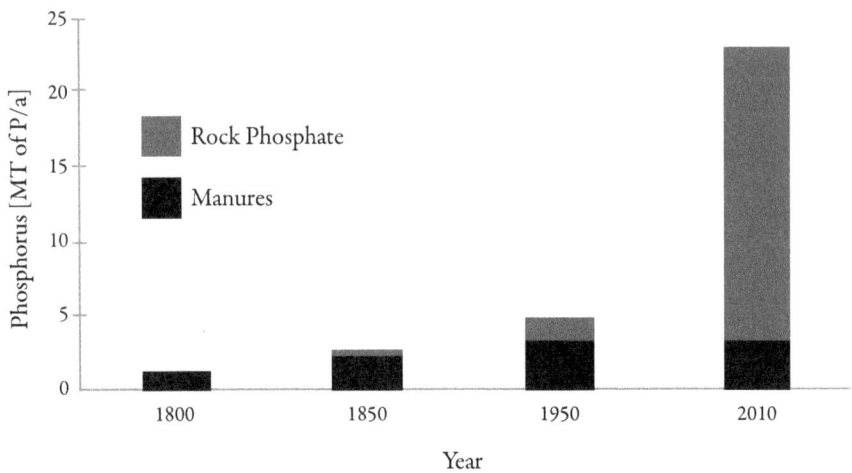

FIGURE 1.4 Phosphorus use and dependency has changed considerably since humans began practicing wide-scale agriculture. Over time, P inputs to field have become far less localized and sourced from increasingly inorganic sources (e.g., mineral forms). Figure created from concepts put forth in Ashley et al. 2011.

large-scale mining, manufacture, and transportation of P fertilizer an economically viable option. As soon as large-volume international trade and growing populations became part of the human P cycle, human P-use became less cyclical and increasingly unsustainable.

Human demand for P is only expected to grow (Childers et al. 2011). If the human population grows by two to three billion people by 2050, and if that larger population is more affluent (Childers et al. 2011), global food production will have to increase by 70 percent to 100 percent (not including added agricultural demand from biofuels production; see chapter 4). In a future in which mineral P resources may become more scarce and expensive, the implications of this dramatic increase in demand are significant for global food security and international relations (chapter 8). Intergenerational equity will also be a growing concern if we continue to manage P resources as we have in the recent past. Clearly a new, more sustainable approach is needed.

HOW A SUSTAINABILITY "LENS" SHAPES SOLUTIONS TO P MANAGEMENT

Sustainability is not simply about using resources more efficiently. In fact, efficiency in food production, including the use of P, has increased through the Green Revolution (Baker et al. 2011), but food security and P sustainability continue to be problems. Thus, although more efficient use of resources may be part of a sustainable solution, a more holistic, or systems, perspective that considers the complex relationships among resources and people is essential. Sustainability focuses on social equity and standard of living, and our relationship to the environment is key because it supports all human activities. Sustainability is not about maintaining the status quo (i.e., sustaining what we currently have). There are plenty of undesirable system-states that are resilient to change (e.g., the "poverty trap"). To the contrary, sustainable solutions seek to improve the current state of a system by accepting our dependence on the environment as our life support system while at the same time including humans and social equity as a fundamental part of the solution. Sustainability is a process, not an endpoint, and most sustainability plans are action-oriented, iterative processes (Figure 1.1, explored in more detail in chapter 9). Effective sustainable solutions require nimble and reflexive responses to changes in the situation. Finally, sustainability science focuses on understanding the trade-offs that inevitably occur when we make various decisions or implement various interventions.

THE TIMING OF P SUSTAINABILITY

Phosphorus resource management is a global sustainability challenge to which there are many possible solutions. Everyone, from households to nations, has a stake in food security, clean water, and geopolitical stability. There is no biological substitute for P, but there are substitutes for the inefficient ways in how we currently use P. Unlike energy, which can only be transformed, P can be recycled. In nature, the P cycle is one of the most conservative of the major macronutrients (Chapin et al. 2002). Thus, instead of materials substitution, we may use "process substitution" with P. This is when we replace inefficient processes that cause P to be lost from the human P cycle with processes that recycle P back into food production. If we were to replace the [largely] linear path of P through our food system with a more cyclical one (Childers et al. 2011), we could dramatically reduce the need for mineral P resources. This "de-linearization" of the P cycle will require a **transdisciplinary*** approach (chapter 9) in which multiple disciplines of science work together and which includes policy makers and practitioners that affect or are affected by a problem (we address the identification of stakeholders in chapter 9). This trans-disciplinary approach allows for a more complete understanding of the problem. In addition, by having multiple interests and perspectives at the table, there is a better chance that negative trade-offs between proposed solutions may be avoided. Perhaps most importantly, this broad approach to identification of challenges and the development of solutions is preferable to having a small group of stakeholders decide what should or should not be done. In other words, holistic approaches are based on "buy-in" by all involved. Together, people can create visions of desirable futures and partake in their realization.

SYSTEMS VS. LINEAR THINKING

A "systems-thinking" approach to sustainability involves solutions that coordinate efforts in multiple sectors of a problem. Truly sustainable solutions to P-use will almost certainly include transformational change rather than a mere collection of small corrections, or "tweaks," to our current system. In the case of P, this means that simply considering solutions for discrete steps in the human P cycle will be insufficient. The use of more P-efficient crop varieties by farmers would be an important "tweak" to the existing agro-industrial complex, while a large-scale rethinking of when, where, and how crops are grown would be a transformational change. For example, perhaps some food must be produced closer to cities, where urban waste (a "source" of P) can be more easily reincorporated into food production. Perhaps polyculture and the reintroduction of small-scale animal husbandry on crop farms would be appropriate to both minimize P losses to waterways while decreasing the amount of P the farm must import. Importantly, holistic approaches should avoid large trade-offs with other essential resources or services. For example, solutions to the P-use issue that substantially increase energy use (especially from fossil fuels) would in most cases be unsatisfactory.

Another important component of the sustainability "lens" is the development of an iterative and reflexive process that regularly assesses how the system is changing in response to interventions and adjusts solutions appropriately (Figure 1.1b). Increased understanding of the current state better informs interventions, but this is not enough to ensure that long-term goals are met. As we intervene to change a system, we change the nature of that system and must be willing to constantly reassess the path forward. The need for such an iterative structure highlights the need for involvement by policy makers, practitioners, and citizens—as they will be able to report what works and what does not work—and they must be part of a nimble, flexible governance structure (by proposing, voting, and participating; chapter 9).

A wide range of technological, behavioral, and political solutions are available to improve reuse and recovery of P resources (as well as efficiency of P resource use), and all of them will likely contribute to achieving P sustainability and thus food security (Cordell et al. 2011). Although P sustainability is a global challenge, vulnerability to P scarcity and to P eutrophication varies considerably around the globe and tends to be manifest much more locally (chapter 2 for biophysical reasons, chapter 4 for case study examples, chapter 7 for cultural norms, chapter 8 for international aspects). Similarly, solutions often need to be tailored to regional, even local, characteristics. For example, one clear goal should be to decrease per capita consumption by, and increase recycling and efficiencies in, developed countries. This will not be possible, though, if it leads to a (real or perceived) erosion in the standard of living in these societies. At the same time, solutions should not hinder improvements in the quality of life in developing countries.

Thus, the sustainability "lens" is a way to examine systems, identify problems, and come up with long-range solutions. The global P-use issue is a "wicked problem" that can benefit from such a sustainability "lens." This book uses this lens to further characterize the challenges we face with current patterns of human interactions with P and to explore strategies to move toward a more sustainable system (Figure 1.5).

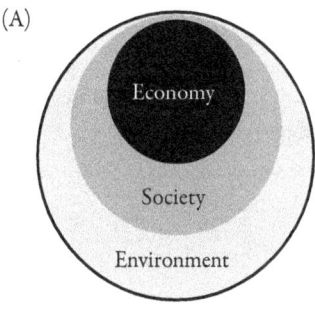

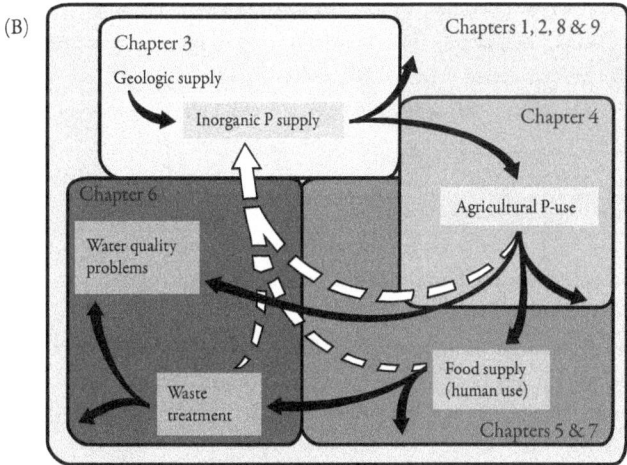

FIGURE 1.5 (A) Each chapter in the book encompasses all three spheres of sustainability. (B) With the human P cycle as a backdrop (modified from Childers et al. 2011), we show the aspects of the cycle that are covered in each of the book chapters. Note that there is considerable overlap among many of the chapters (e.g., 4, 5, 6 and 7) and that chapters 1, 2, 8, and 9 address the entire human P cycle.

> BOX 1.2
> CHAPTER 1 SUMMARY
>
> - Over the course of human history, the human P cycle has changed from a more closed system (recycled and recovered P) to a very open one (P loss) that degrades ecosystem services.
> - P management is a sustainability problem because issues regarding P are integrated and embedded within the three spheres of sustainability (environment, society, and economy).
> - The problem of P sustainability is global, but solutions must be locally appropriate and must consider both time and place.
> - The sustainability lens influences what solutions should be considered in P management by taking a systems perspective rather than a more conventional, linear approach.
> - A sustainability perspective accounts for and integrates environmental, social, and economic influences in a particular time and place. This perspective often involves transdisciplinary engagement in an iterative, problem-solving process working toward a vision of a sustainable future.

REFERENCES

Agyeman, J., R. D. Bullard, and B. Evans. 2003. *Just Sustainabilities: Development in an Unequal World*. The MIT Press, London.

Ashley, K., D. Cordell, D. Mavinic. 2011. A brief history of phosphorus: from the philosopher's stone to nutrient recovery and reuse. *Chemosphere* 84(6):737–46.

Baker, L. A. 2011. Can urban P conservation help to prevent the brown devolution? *Chemosphere* 84(6):779–784.

Bennett, E., S. Carpenter, and N. Caraco. 2001. Human impact on erodable phosphorus and eutrophication: a global perspective. *BioScience* 51:227–234.

Brundtland, H. Our Common Future, Report of the World Commission on Environment and Development, World Commission on Environment and Development, 1987. Published as Annex to General Assembly document A/42/427, Development and International Co-operation: Environment August 2, 1987. Retrieved November 14, 2011.

Carpenter, S. R., N. F. Caraco, D. L. Correll, R. W. Howarth, A. N. Sharpley, and V. H. Smith. 1998. Nonpoint pollution of surface waters with phosphorus and nitrogen. *Ecological Applications* 8:559–568.

Chapin, F., P. Matson, and H. Mooney. 2002. *Principles of Terrestrial Ecosystem Ecology*. Springer-Verlag, New York.

Childers, D. L., J. Corman, M. Edwards, J. J. Elser. 2011. sustainability challenges of phosphorus and food: Solutions from closing the human phosphorus cycle. *BioScience* 61(2):117–123.

Conklin, E. J. 2006. *Dialogue Mapping: Building Shared Understanding of Wicked Problems*. Wiley, Chichester, England; Hoboken, NJ.

Cordell, D. A. 2001. Improving carrying capacity determination: material flux analysis of phosphorus through sustainable aboriginal communities. Thesis, University of New South Wales (UNSW), Sydney, Australia.

Cordell, D., J. O. Drangert, and S. White. 2009. The story of phosphorus: Global food security and food for thought. *Global Environmental Change* 19:292–305.

Cordell, D., A. Rosemarin, J. Schroder, and A. Smit. 2011. Towards global phosphorus security: A systems framework for phosphorus recovery and reuse options. *Chemosphere* 84(6):747–758.

Declaration of the United Nations Conference on the Human Environment. 1972. http://www.unep.org/Documents/Default.asp?DocumentID=97&ArticleID=1503. Accessed September 13, 2011.

Drechsel, P., O. Cofie, M. Fink, G. Danso, F. Zakari, and R. Vasquez. 2004. *"Closing the rural-urban nutrient cycle" Options for municipal waste composting in Ghana.* International Water Management Institute—West Africa.

Dresner, S. 2002. *The principles of sustainability.* Earthscan, London, Sterling, VA.

Falkowski, P., R. J. Scholes, E. Boyle, J. Canadell, D. Canfield, J. Elser, N. Gruber, K. Hibbard, P. Hogberg, S. Linder, F. T. Mackenzie, B. Moore, 3rd, T. Pedersen, Y. Rosenthal, S. Seitzinger, V. Smetacek, and W. Steffen. 2000. The global carbon cycle: a test of our knowledge of earth as a system. *Science* 290:291–296.

Filippelli, G. M. 2008. The global phosphorus cycle: past, present, and future. *Elements* 4:89–95.

Funtowicz, S. O., and J. R. Ravetz. 1993. Science for the post-normal age. *Futures* 25:739–755.

Gibson, R. B. 2006. Sustainability assessment. *Impact Assessment and Project Appraisal* 24:170.

Harding, G. 1968. The tragedy of the commons. *Science* 162:1243–1248.

Kates, R., W. Clark, R. Corell, J. M. Hall, C. C. Jaeger, I. Lowe, J. J. McCarthy, H. J. Schellnhuber, B. Bolin, and N. M. Dickson. 2001. Sustainability science. *Science* 292:641–642.

Liu, Y., G. Villalba, R. U. Ayres, and H. Schroder. 2008. Global phosphorus flows and environmental impacts from a consumption perspective. *Journal of Industrial Ecology* 12:229–247.

Robert, K.-H., B. Schmidt-Bleek, J. A. d. Larderel, G. Basile, J. Jansen, R. Kuehr, P. P. Thomas, M. Suzuki, P. Hawken, and M. Wackernagel. 2002. Strategic sustainable development—selection, design and synergies of applied tools. *Journal of Cleaner Production* 10:194–214.

Sarewitz, D., D. Kriebel, R. Clapp, C. Crumbley, P. Hoppin, M. Jacobs, and J. Tickner. 2010. *The Sustainable Solutions Agenda.* Consortium for Science, Policy and Outcomes, Arizona State University, Tempe, AZ.

Scholz, R. W., D. J. Lang, A. Wiek, A. I. Walter, and M. Stauffacher. 2006. Transdisciplinary case studies as a means of sustainability learning: historical framework and theory. *International Journal of Sustainability in Higher Education* 7:226–251.

Tepper, Y., 2007. Soil improvement and agricultural pesticides in antiquity: examples from archaeological research in Israel. In *31st Proceedings Middle East Gardens Traditions: Unity and Diversity. Dumbarton Oaks Colloquium on the History of Landscape Architecture*, Harvard University Press, Washington, DC.

Villalba, G., Y. Liu, H. Schroder, and R. U. Ayres. 2008. Global phosphorus flows in the industrial economy from a production perspective. *Journal of Industrial Ecology* 12:557–569.

Van Wambeke, A., F. Rom, and F. Land. 2004. Soils of the tropics. *Geoderma* 123:373–375.

2

The Biology and Ecology of Phosphorus in Biota, Natural Ecosystems, and Agroecosystems

P IS FOR PHOSPHORUS

James J. Elser, William M. Roberts, Philip M. Haygarth

Ericka Cero Wood
Lake Atitlán, then and now

Scientific Collaborator:
Jessica Corman, PhD student,
School of Life Sciences, Arizona State University

Description of Artwork:
My scientific collaboration was with Jess Corman, a PhD student in the School of Life Sciences at ASU. Her studies focus on how elements (including phosphorus) cycle in ecosystems. Jess is involved with researching the algae blooms at Lake Atitlán in Guatemala and the potential link to phosphorus in agricultural runoff and untreated human wastewater discharge. Lake Atitlán is the source of fresh water for 400,000 people, yet currently there is little to no solid waste treatment facilities and waste is instead released directly into the lake. Ideas to address this issue range from waste treatment plants, to natural wetlands, and new technology such as wastewater ash fertilizer. With the use of nitrogen and phosphorus in commercial fertilizers, agricultural pollution has become a major issue for all fresh water, and Lake Atitlán is no

exception. When water containing an excess of nitrogen and phosphorus is coupled with a warming climate, it creates perfect conditions for the blooms of cyanobacteria that deplete oxygen, killing fish and creating a stench of rotting plant material. Lake Atitlán has encountered ecological turmoil for over 60 years, including the introduction of non-native black bass that decimated native fish and led to the extinction of the giant grebe bird. Algae blooms may become a permanent part of Atitlán unless something is done to control pollution entering the lake. The blooms have become so large that, as shown by NASA satellite images taken in November 2009, the entire 12-mile-long, quarter-mile-deep lake is covered in pea-green algae. Phosphorus is needed by all living things, but too much of it is toxic. What is necessary is that there be a balance. The algae blooms are telling us that things are not right. My art depicts Lake Atitlán as it was, and how it may be in the future, with phosphorus now a part of the water forever.

About the Artist:
"Blow it up! Make it bigger! Even small things are beautiful. You don't have to paint the whole thing." These were the words from Ericka's high school art teacher, and they have inspired her to capture the details that create the big picture. Always passionate about the earth and its web of life, Ericka looks forward to the day when environmental protection includes not only the "wilds," but also the urban places where we live and work. Ericka's clients include the Maricopa County Libraries Summer Reading Program, Sedona Jazz on the Rocks, Best Western, and numerous private collectors.

Ericka Cero Wood; *Lake Atitlán, then and now;* acrylic/mixed media, 16" x 20", 2011.

> BOX 2.1
> CHAPTER 2 OBJECTIVES
>
> - Describe how P is used in biological molecules, structures, and physiology.
> - Illustrate how P acts as a limiting nutrient for production and growth for organisms and ecosystems.
> - Explain how P cycles in natural and agricultural ecosystems and causes aquatic eutrophication.

PHOSPHORUS IS CENTRAL IN BIOLOGY

First purified from the urine of German soldiers by seventeenth-century alchemist Hennig Brand, phosphorus (P) has 15 protons in its atomic nucleus and thus is the 15th element in the Periodic Table. Note that in this chapter and in the rest of the book we will refer repeatedly to "phosphorus," even though little phosphorus exists in pure elemental form under normal Earth conditions. This is because its outer electron orbital contains only five electrons out of a capacity for 18. Thus, P atoms can readily react to share electrons with other atoms and especially those of oxygen (O), which themselves are highly reactive. As a result, nearly all P atoms are bound to four oxygen atoms in the form of **phosphate,*** PO_4^{3-}; PO_4, hereafter.) While it may seem obvious to many, we think it is important to note, from the outset, the fundamental fact that P is a chemical **element.*** This is for two reasons. Like most chemical elements, we cannot manufacture P (short of creating the equivalent of a new star to support a nuclear fusion cascade, the original source of all of Earth's P). This is "bad news" from the perspective of potential issues related to P scarcity. However, it is also "good news," because, while the fact that P is a chemical element means that we cannot create any more, it also means that we have not "lost" any P either. That is, all of the P atoms that were on Earth at the start of industrial agriculture are still here (somewhere). Likewise, all of the P atoms that will be used to support agriculture in the future will not "go away" and will still exist. Thus, as is explored in the rest of this book, it is the *distribution* and *availability* of these P atoms that will be of central importance, as they shift from highly concentrated "stockpiles" (e.g., phosphorite deposits) that are cheap to extract (at least for those in affluent countries) to highly diffuse and distributed sources that may be quite expensive to recover (e.g., from riverine, lacustrine, or marine sediments).

So, why is P so central to agricultural productivity and to human well-being (Figure 2.1)? The answer lies in P's central role in nearly all of the key biochemical functions in every living thing. Perhaps P's most important function in a cell is as a key structural component of nucleic acids. P "holds our genes up." That is, each nucleotide

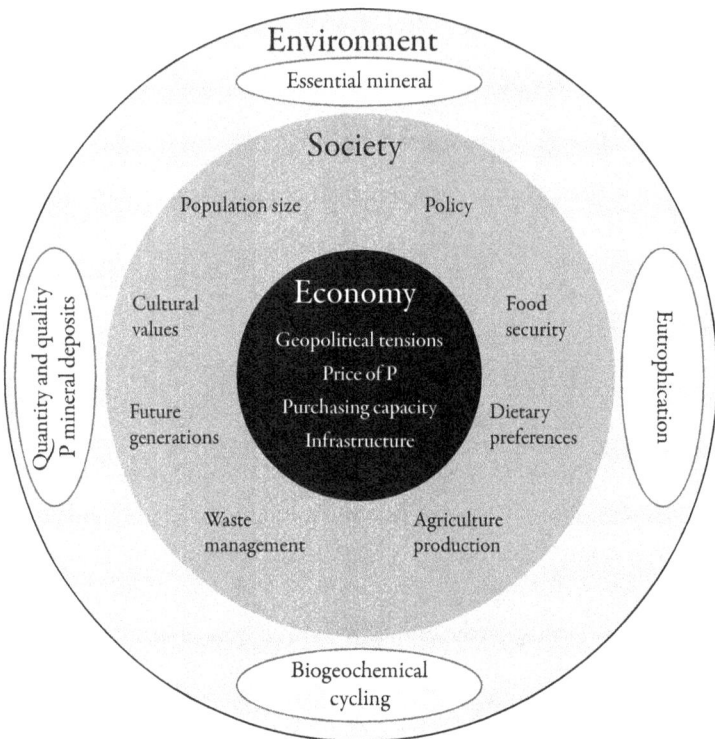

FIGURE 2.1 Biological demand for phosphorus works in all three spheres of sustainability thinking. Phosphorus plays a large role in environment (molecule construction), society (agriculture and diet), and the economy (eutrophication).

building block (e.g., A, T (U), C, G) of the nucleic acids **DNA*** and **RNA*** contains a phosphate (PO_4) unit that provides the bridging link to the next nucleotide in a nucleic acid polymer (Figure 2.2). The contribution is not trivial: P atoms constitute ~9 percent of the total mass of a nucleic acid (both DNA and RNA; Sterner and Elser 2002). In most cells, the dominant nucleic acid is not DNA (which tends to get all the publicity) but instead is RNA; RNA:DNA mass ratios often range from 3:1 to as high as 20:1 (Sterner and Elser 2002). In most cells, the dominant form of RNA is ribosomal RNA (rRNA), which makes up >85 percent of total cellular RNA in most situations. Along with ribosomal proteins, this rRNA is used in construction of the key organelle of protein synthesis, the ribosome. Importantly, rapidly growing cells tend to have increased allocation to ribosomes and thus rRNA and thus P, a set of connections known as the "Growth Rate Hypothesis" (Elser et al. 1996, 2000). While this hypothesis is most strongly supported in observations of heterotrophic bacteria and small metazoans (Elser et al. 2003), for which rRNA contribute 30–80 percent of total P content, it may also be applicable to photosynthetic microbes (Loladze and Elser 2011) and plants (Niklas 2006, Reich et al. 2010; but see Matzek and Vitousek 2009). Indeed, genetic studies of crop plants suggest that selection for increased yield led to increased capacity for ribosome production (Elser et al. 2000), implying that high-yield crop varieties might have intrinsically high P requirements due to the high P demands of ribosome production. A key implication of this is that there might be an unavoidable trade-off between high yield and P requirement, complicating efforts to develop a more P-efficient agricultural system. To be clear: what this implies is that attempts to develop a more P-efficient crop plant that is also agronomically viable may be doomed to failure. Since rapid growth and production generally means investment in more (P-rich) ribosomes, such a low-P plant would be inherently slow growing. Perhaps the answer is to develop (somehow) better ribosomes that, for example, have higher protein synthesis rates per unit. However, given that evolution has been operating on the structure and function of ribosomes for billions of years, this prospect seems daunting.

The importance of P extends beyond its key role in growth-related nucleic acid allocation. For example, nearly all cell membranes are constructed using P-containing phospholipids (Figure 2.2); the phosphate group has a major effect on the physical and chemical properties of the membrane. The P content of a typical phospholipid is ~4 percent, but the overall contribution of phospholipids in most organisms is relatively small (<6 percent of dry mass, Sterner and Elser 2002). Phosphorus is also centrally involved in energy

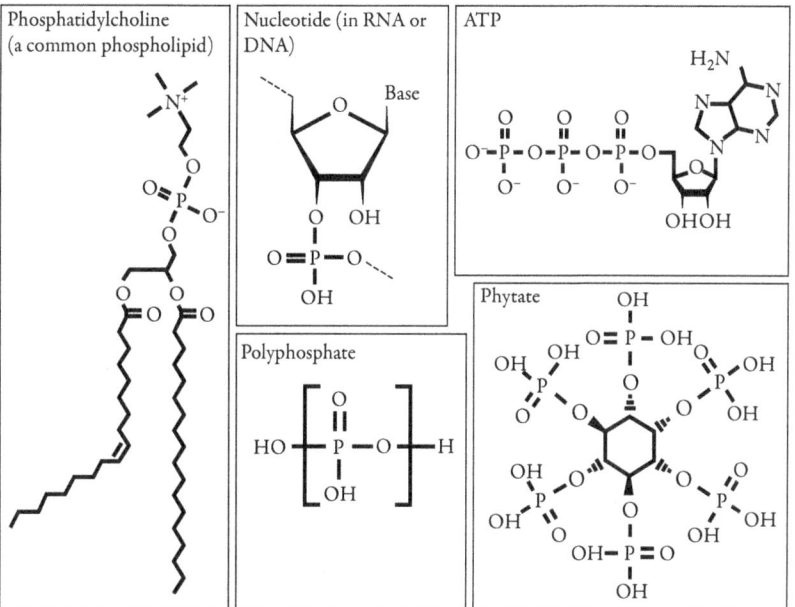

FIGURE 2.2 Common biological molecules that contain P. In many organisms, nucleic acids (especially RNA) are the dominant pool. Under certain conditions bacteria can store P as polyphosphate while many plants, including crops, provision P to their seeds via phytate, a form of P that is indigestible for non-ruminant animals.

metabolism, as the "key currency" for energy transactions is phosphate–phosphate bonding in ATP (Figure 2.2). ATP is a very P-rich molecule (~18 percent P, Sterner and Elser 2002). However, these energetic nucleotides contribute very little to total biomass (<0.1 percent of dry mass) and so do not strongly affect an organism's P requirement.

Three more forms of P are of particular relevance to agricultural P-use and the human P system. In many plants, excess supplies of nutrients can be stored for later use or, in the case of vascular plants, provisioned to seeds for use by the developing sprout. An important molecule for P storage and provisioning is **phytate*** (Figure 2.2), which is present in high concentrations in various seeds and grains (especially corn). Importantly, phytate P is not biologically available to non-ruminant animals because non-ruminants cannot produce the enzymes needed to break the phytate ring structure. Thus the efficiency of P-use from feed can be very low for such livestock (e.g., pigs, but see the Enviropig™ described in chapter 4). In microbes, excess P (and energy) can be stored in long polymers of phosphate called **polyphosphates*** (Figure 2.2). These can be especially important forms of P for microbes in wastewater treatment plants, where both organic energy and P are in great abundance leading to polyphosphate accumulation in microbial cells and thus more efficient retention of P during the wastewater treatment process (see chapter 6; Oehmen et al. 2007). Finally, vertebrate animals (e.g., mammals, birds, fish, etc.) contain considerable quantities of P in the form of the mineral **apatite*** (hydroxyapatite: $Ca_{10}(PO_4)_6(OH)_2$) in bones and teeth. In most vertebrates, >90 percent of total body P is in the bones at any given time (Sterner and Elser 2002) and therefore P, along with Ca, is an important component of proper mineral nutrition in livestock and poultry rearing and in fish aquaculture (see Figure 2.3C). For the same reasons, P is also essential for humans in building and maintaining healthy bones.

The key bottom line, then, from this consideration of the role of P in molecules and organisms is the following: *there is no substitute for P in biology.* So, given its essential nature, how do organisms obtain P?

Plants acquire essentially all of their P via their roots and take it up almost entirely in the form of PO_4 present in soil solution; in many cases, this P acquisition is done in association with fungal symbionts (see next paragraph for further discussion of fungal symbionts). Generally P is transported into the roots via specialized phosphate transporters, where it becomes involved directly in root metabolic processes, is stored, or is transported through the root system to above-ground parts of the plant. Plants can respond to changes in soil P supply levels through a variety of mechanisms that increase their ability to acquire P (Vance et al. 2003). These include changes in root architecture to explore more soil volume (especially surface soil layers), increases in the density and affinity of PO_4 transporters to acquire PO_4 at low soil solute concentrations, release of organic acids into the rhizosphere to liberate inorganically bound forms of P, and production of enzymes such as phosphatases to release organically bound P. Crop

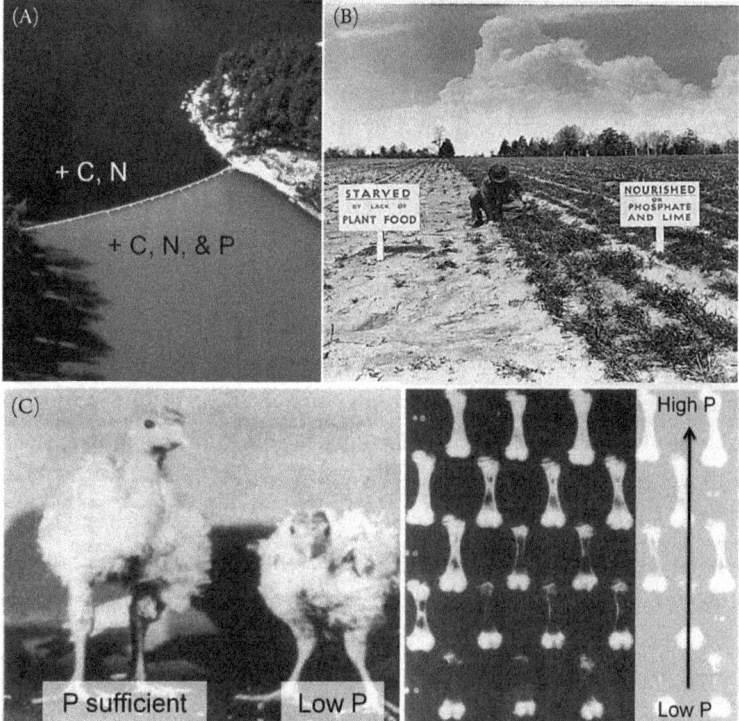

FIGURE 2.3 Phosphorus is limiting for ecosystems and organisms. A. Experimental Lake 226 at the Experimental Lakes Area (Ontario, Canada) after experimental fertilization (Schindler 1977). The top half of the basin received organic carbon and nitrogen (as NH_4 and NO_3), while the lower half received the same inputs of C and N along with P. (Photo courtesy of E. DeBruyn/Fisheries and Oceans Canada) B. Clear effects of P fertilization are shown in this archival photograph from the Tennessee Valley Authority. (*Source:* U.S. National Archives and Records Administration ARC identifier 195896) C. Visible effects of dietary P content on development of chickens (left; Photo courtesy of G. D. Butcher, University of Florida) and on bone mineralization in pigs (right; as seen in X-rays; photo courtesy of GE Combs, Cornell University), (from McDowell 2000).

plants have many of these same talents in their physiological repertoires, but often to a greatly reduced degree. Thus, strategies to improve crop P-use efficiency are aimed at improving crop abilities in these areas; some of these will be discussed in chapter 4. A plant can also access dissolved soil P directly when it moves water into roots due to transpiration. However, this process generally accounts for little of the overall P demand of plants, even in P-rich agricultural soils (Chapin et al. 2002).

Fungi play a major role in soil P cycling and often mediate the acquisition of P by vascular plants via symbiotic associations called mycorrhizae. Such symbioses are especially important in P-deficient soils as plants subsidize their fungal symbionts with fixed C translocated to the roots while the mycorrhizae provide the plant with P. The result of the symbiosis is generally that the combined root/mycorrhizal system has considerably more net surface area and can much more efficiently extract soil P than plant roots without mycorrhizae. These interactions are important in the context of agricultural P sustainability, as most tilling practices in industrial agriculture disrupt soil mycorrhizae and thus crops are less efficient at using soil P. Indeed, supplementation of soils with cultured mycorrhizae shows promise in enhancing crop yields in P-deficient soils (Phosri et al. 2010).

Bacteria, including those in soil, can access both inorganic (e.g., PO_4) and organically bound forms of P. As with plants, phosphate-P is moved into the cell via specialized active transport sites. Often, microbes possess genes for more than one kind of transporter and induce the production of high-affinity transporters when P becomes scarce in the environment. Bacteria also mobilize P contained in organic compounds (breaking phospho-ester bonds, such as those in DNA and RNA) by production of phosphatase enzymes that are also generally up-regulated when P is limiting. Other enzymes used to degrade organic matter can also liberate organically bound P. When P supply is sufficient to satisfy bacterial demands, then bacteria can be considered "mineralizers." However, very commonly P is limiting to microbial growth too and thus P mineralization also occurs when bacteria themselves are consumed by soil protozoa and other bactivores (e.g., nematodes) in the soil food web.

In contrast to plants and microbes that can obtain P as PO_4 directly from the external environment, animals must obtain P from their diet, digesting ingested biomass and transporting P, generally in phosphate (PO_4) form, across the gut into the bloodstream, where it is regulated along with Ca in a relatively complex feedback system. This relatively simple picture overlooks important complexities relevant to the P sustainability arena. For example, as mentioned above, many plants, including important cereal crops like corn, store P in seeds in the form of "phytate" (also known as phytic acid; see Figure 2.2), in which six phosphate molecules are bonded to a central 6-carbon alcohol (inositol). Importantly, P in this form is not available to non-ruminant animals (including humans) but can be accessed by ruminants because their gut microbes produce phytase enzymes that can liberate the phosphate. As a result, P availability to non-ruminant livestock can be very low, resulting in both low assimilation efficiencies that demand P supplements in feed as well as in high concentrations of P in manure (Yano et al. 1999). Farmers can offset this by adding phytase to feed, but, of course, this can be expensive. In chapter 4, an approach using genetically engineered pigs that can produce their own phytase is discussed.

PHOSPHORUS IS ESSENTIAL FOR HIGH PRODUCTIVITY

Basic Features of the P Cycle

Unlike the cycles of other major elements such as carbon (C) and nitrogen (N), the global phosphorus cycle lacks a significant gaseous component and thus P cycles predominantly between rock and aqueous phases with tight, biotically driven, recycling loops within ecosystems (Schlesinger 1991). In the absence of human mobilization of P via mining, P is supplied to the global system by relatively slow weathering of calcium phosphate minerals (see chapter 3). This weathering is greatly amplified by biological activities, especially the production of organic acids by microbes and by plant roots. Once entrained into biological systems, P cycles tightly between living and non-living aqueous forms before eventually being transported by streams and rivers for ultimate burial in floodplain, lake, and ocean sediments. Eventually (after hundreds of millions of years), these sedimentary deposits, having been transformed into sedimentary rocks, are uplifted and subjected to weathering, completing the cycle. This brief description highlights the fundamental way that human activities have amplified and linearized this cycle. That is, the "human P cycle" (*sensu* Childers et al. 2011) in modern times is not actually a cycle; instead, industrial agriculture sends P on a largely "one-way trip" that mobilizes P via mining and then processes it into fertilizer for application to fields. As discussed further below, much of the P added to soil is lost via transport in erosion and runoff, moving downstream toward various receiving waters. As for the P removed at harvest, only a small percentage finds its way back to the field (Haygarth et al. 2005), the rest ending up as crop, food, animal, and human wastes that are ultimately buried in landfills or discharged into streams and rivers. This unsustainable scenario is the core subject of this book.

Role of P in Limiting Production

Relatively slow geochemical supplies rates of P from weathering and transport, coupled to the high biological demands for P described earlier, mean that often P supplies are insufficient to meet the demands of primary producers (plants, algae, and cyanobacteria) in many natural ecosystems. Indeed, a global analysis of more than 1000 field experiments that manipulated nitrogen (N) and P supplies, either separately or in combination, demonstrated that the stimulatory effects of P enrichment (added alone) on primary producers are widespread and similar in magnitude across diverse marine, freshwater, and terrestrial ecosystems (Elser et al. 2007). Furthermore, the response to P is similar in magnitude to the effects of N enrichment in both freshwater and terrestrial ecosystems, a potentially surprising result given widespread focus on N limitation in terrestrial ecology. The important role of P in producing increased biomass of algae and cyanobacteria is long-documented for lakes (see Figure 2.3A; Schindler 1977, 1978); recognition of this impact resulted in considerable efforts in controlling point sources of P from wastewater and detergents (Edmondson 1996). While there is ongoing debate about the primacy of P (relative to N) in driving aquatic **eutrophication*** (Lewis and Wurtsbaugh 2008,

Schindler et al. 2008, Conley et al. 2009), what is clear from past research is that elevated inputs of P are necessary (though perhaps not always sufficient on their own) to produce major algal blooms in receiving waters. More discussion of impacts of P loading on ecosystems will come later in this chapter, as well as in other chapters of this book.

Of course, P is also often limiting to production and growth of crop plants (hence the need for added fertilizer; see Figure 2.3B). In fact, it has been estimated that crop yields are limited by soil P in 30–40 percent of arable lands globally (Runge-Metzger 1995, Vonuexkull and Mutert 1995). While the visual symptoms of P deficiency in crop plants can be relatively subtle (e.g., purpling of stems, petioles, and other plant parts), P limitation has major physiological effects, including reduced plant height and overall biomass ("dwarfing" or "stunting"), inhibition of flowering and fruit set, and, of course, ultimately, diminished yield (Figure 2.3B). Sufficient dietary P content is also essential for satisfactory livestock and poultry production, as P deficiency leads to developmental abnormalities, reduced growth, and weak bone formation (Figure 2.3C).

Soil P Cycling

Phosphorus in soil exists in a continuum of organic (that is, involving biological molecules that have carbon–carbon bonds) and inorganic forms (Figure 2.4). These forms exist in different pools within the soil, ranging from the soil solution (which is the pool from which plants actively take up P) to very stable or unavailable pools of inorganic and organic P. Under natural conditions, the main source of P for plant uptake is from chemical dissolution of P from primary minerals (over 150 forms; see chapter 3) contained within the parent soil. However, fluxes derived from this source of P cannot sustain agricultural production at high levels, so inorganic and organic fertilizer is added.

In most temperate soils, 50 to 75 percent of the total phosphorus is inorganic P; however, this soil P fraction can be much smaller in tropical soils (Turner and Engelbrecht 2010). This prevalence of inorganic P forms in soils, especially highly weathered ones, is the outcome of various chemical reactions that geochemically "trap," or fix, P, with a resulting reduction in soil solution P. Geochemical fixation processes include reactions of sorption, precipitation, and occlusion. Precipitation and **occlusion*** are the longer-term reactions that increase secondary mineral pools, resulting in fixation of P in soils (Frossard et al. 1995). In labile inorganic P pools, phosphate is loosely sorbed to soil minerals via a reversible ligand exchange reaction. This more loosely bound pool is more important in maintaining soil solution P than primary and secondary mineral pools. Soil pH determines whether P is sorbed to Ca and Mg or Al and Fe. Chapter 4 discusses genetic engineering approaches to

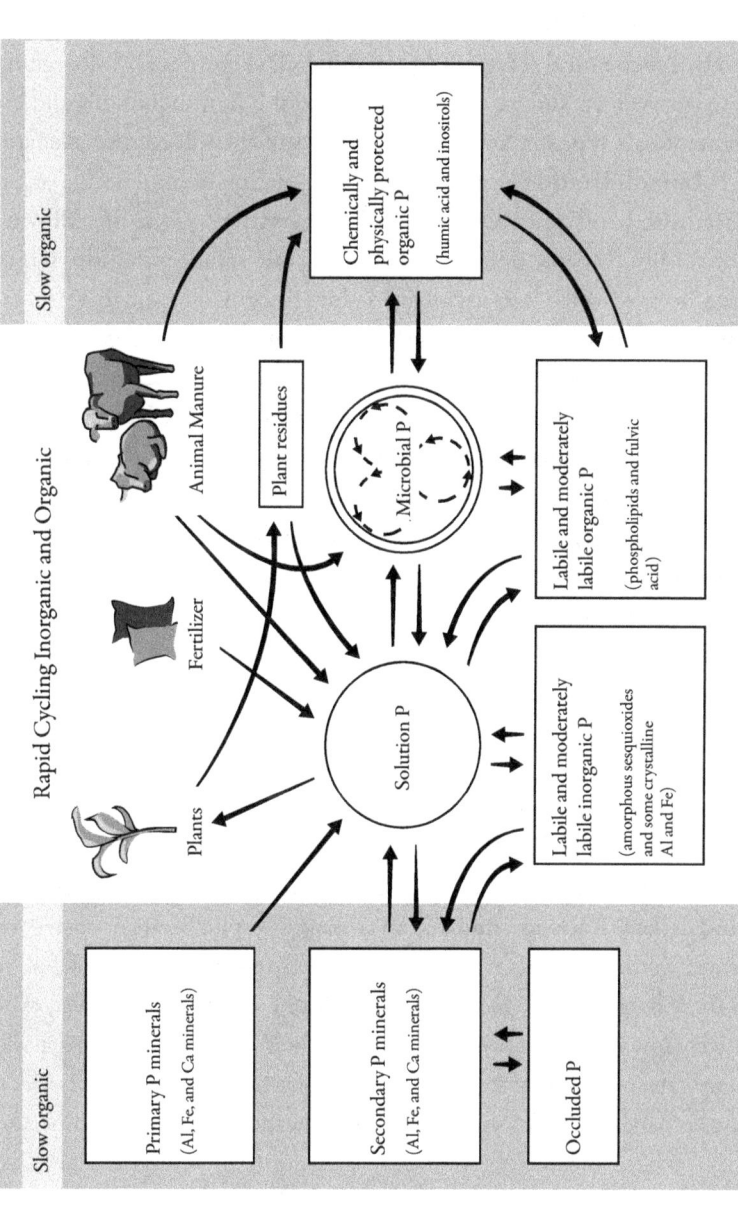

FIGURE 2.4 The phosphorus cycle in agricultural soil (modified from Stewart and Sharpley 1987). The diagram shows that plant (crop) uptake of PO_4 comes from P in solution that itself is supplied by (slow) weathering of P-bearing minerals; by application of fertilizer or manure; by release of P from microbial metabolism or plant biomass; by exchange with labile forms of inorganic P, or by mineralization of organic P forms. The diagram also illustrates that different pools of P have different timescales of turnover and availability. Importantly, many of these pools (such as occluded P and "protected" P) are largely unavailable to plants under normal soil conditions and thus much added fertilizer can accumulate in soil without affecting crop yield.

producing crop plants better able to access these more tightly bound forms of P in soil by acidifying the rhizosphere.

Organic P compounds range from readily available P contained within undecomposed plant residues and soil microbes, to much more stable compounds that have become part of the recalcitrant soil organic matter or are themselves strongly fixed by geochemical reactions to soil minerals (Figure 2.4). Soil organic P can constitute between 30 and 70 percent of total soil P in mineral soils and even more in organic soils. Opposing biological processes in the soil control the fate of this organic P. **Immobilization*** is the microbial formation of more stable organic P that is resistant to breakdown over time. **Mineralization*** is the breakdown or conversion of readily available organic P into inorganic solution P; although this process occurs in most soils, it is usually too slow to provide enough P for crop growth (Figure 2.4). However, because of the large contribution of organic P to the total soil P stock, there is potential to better utilize this P pool to replace or supplement fertilizer addition. Indeed, active efforts are underway to develop approaches to better manage the soil microbial communities to enhance the availability of P to plants (Richardson 2001) and thus to tighten the agricultural P cycle. The fascinating thing is that such approaches, along with those that improve access to inorganically bound forms, mean that nearly all soil P is *potentially* available for uptake given the right soil conditions, timescales, and the biota's need for P. The potential to tap this soil P "resource base" is discussed further in chapter 3.

Agricultural P Transfer through Catchments

As past experiments have shown that over-application of P fertilizer does not reduce crop yield, farmers often apply fertilizer in excess of plant requirements, as a sort of "fertility insurance." For example, from calculated farm budgets it is estimated that grass and arable land in the UK has accumulated an average P surplus of ~1000 kg ha^{-1} over the last 65 years (Withers et al. 2001). Economic losses to the farmer due to P leaching are small, so there has been little motivation to reduce P additions. However, export of even relatively low concentrations of P to freshwater can adversely impact water quality. To assess the potential of such P loss from soils, biogeochemists consider the concentrations of labile forms of soil P assessed via particular extraction procedures (e.g., sodium hydrogen carbonate (NaHCO$_3$) extraction, or "Olsen P") and assess their correlation with P concentrations in surface and subsurface flows (Pote et al. 1996, McDowell and Sharpley 2001). Results of such analyses suggest that large areas of agricultural land pose a risk of P leaching in excess of concentrations likely to have adverse effects on surface waters. This is the case right now, even though the extent of excess P applied to agricultural land is falling in many areas due to P removal in increasing crop yields as well as economic and regulatory drivers for reductions in P fertilizer use (Ulén et al. 2007). Thus, P loss from agriculture contributes a significant proportion of total P (TP) loadings to surface waters and coastal zones (for Europe, 30–50 percent in the year 2000; OECD 2001).

So we have established that heavily fertilized agricultural soils represent an increased risk of P leaching and contribute significantly to P inputs to surface waters, but how does this transfer occur? And what are the consequences of this transfer?

Phosphorus transfer involves several steps that have been called the "**phosphorus transfer continuum**"* (Haygarth et al. 2005; Figure 2.5). Recognizing the steps in this continuum is important in identifying the places where we might implement measures to achieve greater P sustainability. First, there must be a *source* of P (the "source tier"); this could be fertilizers applied to the soil, manure produced by livestock, or P already held in the soils. This source must then be made available for transfer, or *mobilized*. The two main processes involved in mobilization are solubilization and detachment. Solubilization involves geochemical and biological processes such as desorption and enzyme hydrolysis and is therefore closely coupled to soil P cycling. Detachment involves physical processes—for example, surface soil disturbance by heavy rain. To reach surface waters from the point of mobilization, P must be *delivered*. Delivery is dependent on hydrologic processes, as water is the main carrier for P. Delivery may include surface and/or subsurface pathways that vary spatially and temporally. For example, when the soil is saturated or rainfall intensity exceeds infiltration into the soil, P-containing water may flow across the land surface. On the other hand, groundwater or shallow subsurface flow may facilitate P delivery in well-drained soils or during drier periods, but occurs over much longer timescales. Overall, phosphorus lost from agricultural lands as surface runoff is generally greater than the amounts that are lost via subsurface flow. P can also be transferred to water attached to particles and colloids that become physically detached from the soil, most commonly synonymous with soil erosion. It is important to note that additional phosphorus is exported from the landscape via plant harvesting; this component of P export is very significant and varies strongly with the identity of the crop (see chapter 4).

The magnitude of agricultural P loss from leaching, runoff, and erosion across the transfer continuum depends on multiple factors: the rate, timing, and method of P application; the form of fertilizer or manure applied; and the amount and timing of rainfall after application (Sharpley and Rekolainen 1997). Overall, this P loss is considered to be agronomically small (generally <2 kg P ha^{-1}), representing a minor proportion of P applied as fertilizer or manure (generally <5 percent). However, this farm-derived P can nonetheless contribute significantly to *impacts* (the last part of the continuum); the eutrophication of downstream surface waters

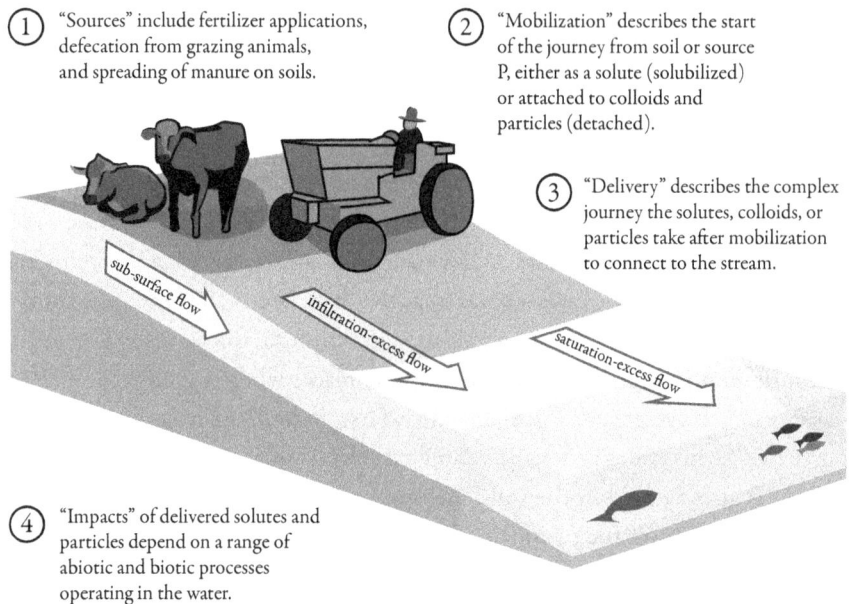

① "Sources" include fertilizer applications, defecation from grazing animals, and spreading of manure on soils.

② "Mobilization" describes the start of the journey from soil or source P, either as a solute (solubilized) or attached to colloids and particles (detached).

③ "Delivery" describes the complex journey the solutes, colloids, or particles take after mobilization to connect to the stream.

④ "Impacts" of delivered solutes and particles depend on a range of abiotic and biotic processes operating in the water.

FIGURE 2.5 The P transfer continuum (Haygarth et al. 2005) depicts the steps by which phosphorus is introduced, mobilized, and transported through agricultural watersheds to have impacts on downstream ecosystems. See text for detailed discussion.

(see Figure 2.3A above) provides a view of what a heavily eutrophied lake can look like. Delivery of total P to surface waters in concentrations exceeding 0.1 mg L^{-1} (or perhaps even lower) will likely impair water quality (Withers and Sharpley 1995), since algal growth in freshwater aquatic systems is often limited by P (see above sections). In summary, the P "transfer continuum" (Figure 2.5) describes source, mobilization, delivery, and impact of P transfer across landscapes, breaking the issues down into conceptual areas that can be used for both process and mitigation understanding (Haygarth et al. 2005).

Mitigation of diffuse P transfer is now becoming increasingly important as governing authorities set increasingly stringent water-quality targets. For agriculture, mitigation options for P transfer focus on different aspects of the P transfer continuum (Haygarth et al. 2009). For example, source reduction may tackle the source tier by reducing fertilizer inputs, while constructed wetlands may tackle the delivery aspects of P transfer by retaining P in overland flow. In terms of soil P cycling, through process manipulation, mobilization of P could be reduced by achieving a soil P status that is enriched in plant-available P and depleted in readily leachable P, thus getting more P into the plant and leaving less to be transported. Of course, the efficacy and risks of any mitigation option considered for large-scale implementation must be fully understood to ensure the desired effect is achieved. As will be emphasized in chapter 5, such efforts will also need to be tailored to local conditions.

EFFECTS OF ELEVATED P INPUTS ON AQUATIC ECOSYSTEMS

The end point of the agricultural P transfer continuum is movement of P into rivers, lakes, and coastal oceans. P also moves into receiving waters from human settlements of all sizes; indeed, in most of the world sewage receives essentially no treatment whatsoever (see chapter 7). Even in developed European countries, the percentage of wastewater subjected to advanced wastewater treatment with P removal varies considerably, from >97 percent in the Netherlands to <4 percent in Turkey (OECD 2004). Given the role of P as a key limiting nutrient in aquatic ecosystems and the fact that P export from watersheds in most of the developed world is dramatically higher than in historical times, we should expect major impacts on downstream ecosystems. And, indeed, lakes and coastal oceans are getting green.

Despite some successes in reducing and mitigating P inputs and recovering local water quality (as in the case of Lake Washington, Seattle; Edmondson 1996), large-scale eutrophication of rivers, lakes, and coastal oceans around the world continues to degrade ecosystem services (such as water supply, recreational amenity, and fisheries production). Multiple articles, books, and reports have covered this dimension of P sustainability (Carpenter et al. 1998, Dale 2010, Smith 2003, Smith et al. 2006, State–EPA Nutrient Innovations Task Group 2009, National Research Council 2000, Committee on Twenty-First Century Systems Agriculture 2010, Carpenter and Bennett 2011, others). Here we merely wish to delineate the extent and multidimensional impacts of nutrient P over-enrichment. Increased nutrient loading is associated with coral reef degradation (McCook 1999), triggers blooms of algae that impair drinking water quality (Hitzfeld et al. 2000), is linked to the expanding layers of coastal oceans that are inhospitable to fish and many other forms of marine life due to oxygen depletion (**hypoxic zones,*** also known as "dead zones," Diaz and Rosenberg 2008, Dale 2010), and reduces the amenity value of lakes and coastal areas (Dodds et al. 2009). Attempts have been made to estimate the economic effects, both direct and indirect, of these widespread impacts. Dodds et al. (2009) developed a conservative estimate that nutrient over-enrichment of inland waters in the United States resulted in ~$2.7 billion of economic costs annually (in 2010 U.S. dollars, here and hereafter), reflecting negative impacts on recreational water usage, waterfront real estate values, spending on recovery of threatened and endangered species, and drinking water provision. For England and Wales, similar estimates for inland waters amounted to ~$162–328 million annually (Pretty et al. 2002). Economic impacts of coastal marine eutrophication are also large (see Anderson et al. 2000, Hoagland et al. 2002). Overall, economic costs from harmful algal blooms between 1987 and 2000 in the United States were as high as $104 million. Such blooms resulted in losses to U.S. commercial fisheries of $32 million annually. To our knowledge, no one has yet attempted to estimate the *global* cost of either freshwater or marine eutrophication.

Reversal of human-caused eutrophication will be one of the big payoffs from achieving a sustainable phosphorus system.

BOX 2.2
CHAPTER 2 SUMMARY

- Phosphorus is a chemical element essential for all living things. This is because of its role in the makeup of key biological molecules, such as DNA and RNA, and structures, such as bones. High productivity and rapid growth have especially high demands for P.
- The natural P cycle is largely a geological cycle, lacking an atmospheric phase. Phosphorus is weathered from rocks, used many times by biota, and then trapped into unavailable forms in soil or sedimented into lakes and oceans.
- Because of high biological demand and relatively slow supplies from rock weathering, P is often limiting to production in aquatic and terrestrial ecosystems. Agricultural soils are also often deficient in P, thus the requirement for P fertilizer addition to increase yield.
- In soil, P can be rapidly immobilized by physical/chemical processes. Thus, in agricultural situations P must often be continuously applied to maintain high yields. On farms, less than half of the P added in fertilizer in a given year is captured in the harvested crop. Considerable P is transferred to water via solubilization (leaching) and physical detachment (soil erosion).
- Due to this downstream transport, inputs of fertilizer P from farms, as well as inputs of wastewater P from towns and cities, cause freshwater and marine eutrophication ("algal blooms," "dead zones").

REFERENCES

Anderson, D. M., P. Hoagland, Y. Kaoru, and A. W. White. 2000. *Estimated annual economic impacts from harmful algal blooms (HABs) in the United States.* Woods Hole Oceanographic Institute, Woods Hole, MA.

Carpenter, S. R. and E. M. Bennett. 2011. Reconsideration of the planetary boundary for phosphorus. *Environmental Research Letters* 6:12.

Carpenter, S. R., N. F. Caraco, D. L. Correll, R. W. Howarth, A. N. Sharpley, and V. H. Smith. 1998. Nonpoint pollution of surface waters with phosphorus and nitrogen. *Ecological Applications* 8:559–568.

Chapin, F. S., H. A. Mooney, M. C. Chapin, and P. Matson. 2002. *Principles of Terrestrial Ecosystem Ecology.* Springer, New York.

Childers, D. L., J. Corman, M. Edwards, and J. J. Elser. 2011. Sustainability challenges of phosphorus and food: Solutions from closing the human phosphorus cycle. *BioScience* 61:117–124.

Committee on Twenty-First Century Systems Agriculture. 2010. *Toward Sustainable Agricultural Systems in the 21st Century*. The National Academies Press, Washington, DC.

Conley, D. J., H. W. Paerl, R. W. Howarth, D. F. Boesch, S. P. Seitzinger, K. E. Havens, C. Lancelot, and G. E. Likens. 2009. Eutrophication: time to adjust expectations (response). *Science* 324:724–725.

Dale, V. H., C. Kling, J. L. Meyer, J. Sanders, H. Stallworth, T. Armitage, D. Wangsness, T. S. Bianchi, A. Blumberg, W. Boynton, D. J. Conley, W. Crumpton, M. B. David, D. Gilbert, R. W. Howarth, R. Lowrance, K. Mankin, J. Opaluch, H. Paerl, K. Reckhow, A. N. Sharpley, T. W. Simpson, C. Snyder, and D. Wright. 2010. *Hypoxia in the Northern Gulf of Mexico*. Springer, New York.

Diaz, R. and R. Rosenberg. 2008. Spreading dead zones and consequences for marine ecosystems. *Science* 321:926–928.

Dodds, W. K., W. W. Bouska, J. L. Eitzmann, T. J. Pilger, K. L. Pitts, A. J. Riley, J. T. Schloesser, and D. J. Thornbrugh. 2009. eutrophication of US freshwaters: analysis of potential economic damages. *Environmental Science & Technology* 43:12–19.

Edmondson, W. T. 1996. *The Uses of Ecology: Lake Washington and Beyond*. University of Washington Press, Seattle.

Elser, J. J., K. Acharya, M. Kyle, J. Cotner, W. Makino, T. Markow, T. Watts, S. Hobbie, W. Fagan, J. Schade, J. Hood, and R. W. Sterner. 2003. Growth rate—stoichiometry couplings in diverse biota. *Ecology Letters* 6:936–943.

Elser, J. J., M. E. S. Bracken, E. E. Cleland, D. S. Gruner, W. S. Harpole, J. Hillebrand, J. T. Ngai, E. W. Seabloom, J. B. Shurin, and J. E. Smith. 2007. Global analysis of nitrogen and phosphorus limitation of primary production in freshwater, marine, and terrestrial ecosystems. *Ecology Letters* 10:1135–1142.

Elser, J. J., D. Dobberfuhl, N. A. MacKay, and J. H. Schampel. 1996. Organism size, life history, and N:P stoichiometry: towards a unified view of cellular and ecosystem processes. *BioScience* 46:674–684.

Elser, J. J., R. W. Sterner, E. Gorokhova, W. F. Fagan, T. A. Markow, J. B. Cotner, J. F. Harrison, S. E. Hobbie, G. M. Odell, and L. J. Weider. 2000. Biological stoichiometry from genes to ecosystems. *Ecology Letters* 3:540–550.

Frossard, E., M. Brossard, M. J. Hedley, and A. Metherell. 1995. Reactions controlling the cycling of P in soils. In *Phosphorus in the global environment: transfers, cycles and management*, edited by H. Tiessen. John Wiley and Sons, New York.

Haygarth, P. M., H. ApSimon, M. Betson, D. Harris, R. Hodgkinson, and P. J. A. Withers. 2009. Mitigating diffuse phosphorus transfer from agriculture according to cost and efficiency. *Journal of Environmental Quality* 38:2012–2022.

Haygarth, P. M., L. M. Condron, A. L. Heathwaite, B. L. Turner, and G. P. Harris. 2005. The phosphorus transfer continuum: Linking source to impact with an interdisciplinary and multi-scaled approach. *Science of the Total Environment* 344:5–14.

Hitzfeld, B., S. Hoger, and D. Dietrich. 2000. Cyanobacterial toxins: removal during drinking water treatment, and human risk assessment. *Environmental Health Perspectives* 108:113–122.

Hoagland, P., D. M. Anderson, Y. Kaoru, and A. W. White. 2002. The economic effects of harmful algal blooms in the United States: Estimates, assessment issues, and information needs. *Estuaries* 25:819–837.

Lewis, W. M., Jr. and W. W. Wurtsbaugh. 2008. Control of lacustrine phytoplankton by nutrients: erosion of the phosphorus paradigm. *International Review of Hydrobiology* 93:446–465.

Loladze, I. and J. J. Elser. 2011. The origins of the Redfield nitrogen-to-phosphorus ratio are in a homoeostatic protein-to-RNA ratio. *Ecology Letters* 14:244–250.

Matzek, V. and P. M. Vitousek. 2009. N : P stoichiometry and protein : RNA ratios in vascular plants: an evaluation of the growth-rate hypothesis. *Ecology Letters* 12:765–771.

McCook, L. J. 1999. Macroalgae, nutrients and phase shifts on coral reefs: scientific issues and management consequences for the Great Barrier Reef. *Coral Reefs* 18:357–367.

McDowell, R. W., editor. 2000. *Minerals in Animal and Human Nutrition*. Elsevier, New York.

McDowell, R. W. and A. N. Sharpley. 2001. Approximating phosphorus release from soils to surface runoff and subsurface drainage. *Journal of Environmental Quality* 30:508–520.

National Research Council. 2000. *Clean Coastal Waters: Understanding and Reducing the Effects of Nutrient Pollution Committee on the Causes and Management of Eutrophication*.

Niklas, K. J. 2006. Plant allometry, leaf nitrogen and phosphorus stoichiometry, and interspecific trends in annual growth rates. *Annals of Botany* 97:155–163.

OECD. 2001. *Environmental Indicators for Agriculture*. Organisation for Economic Co-Operation and Development, Paris.

OECD. 2004. *The OECD Environmental Strategy: Progress in Managing Water Resources*. Page 11 *in* OECD, editor. OECD Observer. OECD, Paris.

Oehmen, A., P. C. Lemos, G. Carvalho, Z. G. Yuan, J. Keller, L. L. Blackall, and M. A. M. Reis. 2007. Advances in enhanced biological phosphorus removal: From micro to macro scale. *Water Research* 41:2271–2300.

Phosri, C., A. Rodriguez, I. R. Sanders, and P. Jeffries. 2010. The role of mycorrhizas in more sustainable oil palm cultivation. *Agriculture Ecosystems & Environment* 135:187–193.

Pote, D. H., T. C. Daniel, P. A. Moore, Jr., D. J. Nichols, A. N. Sharpley, and D. R. Edwards. 1996. Relating extractable soil phosphorus to phosphorus losses in runoff. *Soil Science Society of America Journal* 60:855–859.

Pretty, J. N., C. F. Mason, D. B. Nedwell, and R. E. Hine. 2002. *A Preliminary Assessment of the Environmental Costs of the Eutrophication of Fresh Waters in England and Wales*. University of Essex, Colchester, UK.

Reich, P. B., J. Oleksyn, I. J. Wright, K. J. Niklas, L. Hedin, and J. J. Elser. 2010. Evidence of a general 2/3-power leaf nitrogen to phosphorus scaling among major plant groups and biomes. *Philosophical Transactions of the Royal Society B: Biological Sciences* 277:877–883.

Richardson, A. E. 2001. Prospects for using soil microorganisms to improve the acquisition of phosphorus by plants. *Functional Plant Biology* 28:897–906.

Runge-Metzger, A. 1995. Closing the cycle: obstacles to efficient P management for improved global security. In *Phosphorus in the Global Environment*, edited by F. Follet, J. W. B. Stewart, and C. V. Cole, 27–42. John Wiley and Sons Ltd, Chichester, UK.

Schindler, D. W. 1977. Evolution of phosphorus limitation in lakes. *Science* 195:260–262.

Schindler, D. W. 1978. Factors regulating phytoplankton production and standing crop in the world's freshwaters. *Limnology and Oceanography* 23:478–486.

Schindler, D. W., R. E. Hecky, D. L. Findlay, M. P. Stainton, B. R. Parker, M. J. Paterson, K. G. Beaty, M. Lyng, and S. E. M. Kasian. 2008. Eutrophication of lakes cannot be controlled by reducing nitrogen input: Results of a 37-year whole-ecosystem experiment. *Proceedings of the National Academy of Sciences USA* 105:11254–11258.

Schlesinger, W. H. 1991. *Biogeochemistry: An Analysis of Global Change*. Academic Press, San Diego, CA.

Sharpley, A. N. and S. Rekolainen. 1997. Phosphorus in agriculture and its environmental applications. In *Phosphorus Loss from Soil to Water*, edited by H. Tunney, et al. CAB International, Wallinford, UK.

Smith, V. H. 2003. Eutrophication of freshwater and coastal marine ecosystems—A global problem. *Environmental Science and Pollution Research* 10:126–139.

Smith, V. H., S. B. Joye, and R. W. Howarth. 2006. Eutrophication of freshwater and marine ecosystems. *Limnology and Oceanography* 51:351–355.

State–EPA Nutrient Innovations Task Group. 2009. An Urgent Call to Action—Report of the State–EPA Nutrient Innovations Task Group.

Sterner, R. W. and J. J. Elser. 2002. *Ecological Stoichiometry: The Biology of Elements from Molecules to the Biosphere*. Princeton University Press, Princeton, NJ.

Stewart, J. W. B. and A. N. Sharpley. 1987. Controls on dynamics of soil and fertilizer phosphorus and sulfur. In *Soil Fertility and Organic Matter as Critical Components of Production*, edited by R. F. Follet, J. W. B. Stewart, and C. V. Cole, 101–121. American Society of Agronomy, Madison, WI.

Turner, B. L. and B. M. J. Engelbrecht. 2010. Soil organic phosphorus in lowland tropical rain forests. *Biogeochemistry* 103:297–315.

Ulén, B., M. Bechmann, J. Fölster, H. P. Jarvie, and H. Tunney. 2007. Agriculture as a phosphorus source for eutrophication in the north-west European countries, Norway, Sweden, United Kingdom and Ireland: a review. *Soil Use and Management* 23:5–15.

Vance, C. P., C. Uhde-Stone, and D. L. Allan. 2003. Phosphorus acquisition and use: critical adaptations by plants for securing a nonrenewable resource. *New Phytologist* 157:423–447.

Vonuexkull, H. R. and E. Mutert. 1995. Global extent, development, and economic impact of acid soils. *Plant and Soil* 171:1–15.

Withers, P. J. A., A. C. Edwards, and R. H. Foy. 2001. Phosphorus cycling in UK agriculture and implications for phosphorus loss from soil. *Soil Use and Management* 17:139–149.

Withers, P. J. and A. N. Sharpley. 1995. Phosphorus fertilizers. In *Soil Amendments and Environmental Quality, edited by* J. E. Rechcigl, 65–107. CRC Press, Boca Raton, FL.

Yano, F., T. Nakajima, and M. Matsuda. 1999. Reduction of nitrogen and phosphorus from livestock waste: A major priority for intensive animal production—Review. *Asian-Australasian Journal of Animal Sciences* 12:651–656.

3

Global Phosphorus: Geological Sources and Demand-Driven Production

P IS FOR PRICE

Donald Burt, Marion Dumas, Nathaniel Springer, David A. Vaccari

Adam Farcus; *Weight of the World (1 / 3.05E+28 grams);* Phosphorus rock bought on eBay. 5" x .5" x .5", 2010.

Adam Farcus
Weight of the World (1 / 3.05E+28 grams)

Scientific Collaborator:
Rebecca Cors, PhD student, Swiss Federal Institute of Technology

Description of Artwork:
I try to instill in my work the magic and poetry that I find in everyday life. For this piece I decided to represent the urgent need to solve the widely unknown global problem of phosphorus depletion. The fraction in the title is the mass of this piece of phosphorus over the mass of the Earth. By referencing the weight of the world, the piece of phosphorus acts as a metaphor for the importance of phosphorus.

About the Artist:
Adam Farcus currently lives in the East Garfield Park neighborhood of Chicago and has taught at the University of St. Francis since 2009. His work has been exhibited at the Gallery 400, Chicago; University Galleries, Normal, IL; Hyde Part Arts Center, Chicago; Second Bedroom Project Space, Chicago; the Miami Bridge Art Fair, Miami; and the Urban Institute for Contemporary Arts, Grand Rapids; among many others. Adam received his MFA from the University of Illinois at Chicago, his BFA from Illinois State University, and his AA from Joliet Junior College. In the summer of 2010 he co-chaired a panel discussion at the Performance Studies International 16 at OCAD in Toronto, Ontario.

BOX 3.1
CHAPTER 3 OBJECTIVES

- Describe where and in what form phosphorus occurs on Earth.
- Identify where and in what quantities extractable phosphorus occurs.
- Elucidate what the past and current trends of phosphorus usage are and how future trends can be predicted, and with what reliability.
- Explain if scarcity of phosphorus is a potential problem and, if so, in what sense and in what time frame.

PHOSPHORUS RESOURCES

Crustal Occurrence

Phosphorus (P) is a relatively common element (tenth most abundant) in the Earth's crust and can occur in hundreds of distinct phosphate minerals (Huminicki and Hawthorne 2002). The major geological dimensions of phosphorus (Figure 3.1) are

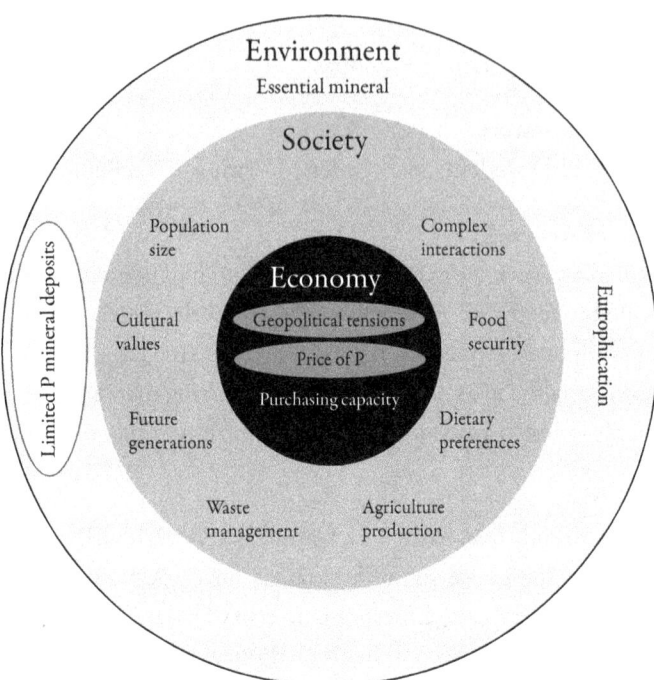

FIGURE 3.1 Phosphorus supply and availability works in all three spheres of sustainability thinking: environment (mining techniques), society (mine safety), and the economy (ore prices).

essential to the development of a comprehensive approach to P sustainability, and it is with this topic that we begin our discussion in this chapter.

Much, and perhaps most, P in rocks occurs in a highly diluted form, by means of the coupled solid solution of P plus aluminum (Al) for two silicons in the rock-forming mineral feldspar, which is the most abundant mineral in the crust (e.g., Manning 2008). Weathering of feldspars to clay-rich soils releases this phosphorus, making it available for plants, though often not rapidly enough to meet plant demand (chapter 2). These fluxes are also insufficient for intensive industrialized agriculture, necessitating the use of mined, soluble phosphorus in the form of fertilizer.

Despite its wide and varied distribution, almost all the P that is worth mining occurs in a single phosphate mineral group, **apatite,*** in two distinct types of deposits, sedimentary and igneous (discussed below in *Exploration and Discovery* section). In greatly simplified form, apatite in nature can be considered as mainly a solid solution between hydroxyl-apatite, or $Ca_5(PO_4)_3(OH)$, and fluor-apatite, or $Ca_5(PO_4)_3(F)$. Note: "solid solution" means that the OH^- and F^- ions substitute freely for each other in the way that metal atoms substitute for each other (or dissolve in each other) in a solid alloy. The apatite crystal structure is subject to multiple other ionic substitutions of ions of similar sizes and charges (Pan and Fleet 2002). This has important implications for P sustainability, because some of these ions, such as Cd (cadmium) and U (uranium), can have deleterious environmental consequences, even when present in only minor amounts (e.g., Manning 2008). Note: the nomenclature of the apatite mineral group was revised in 2007 by a subcommittee of the International Mineralogical Association (IMA) so as to make the two minerals above apatite-(CaOH) and apatite-(CaF), respectively (Burke 2008), but this nomenclature has not yet found wide acceptance.

Apatite is a common mineral not only because it is made of common elements, calcium and phosphorus, but also because it is the most stable (least water soluble) common natural phosphate. This stability, plus relative physical toughness, makes apatite biologically suitable for constructing vertebrate bones and teeth (Pasteris et al. 2008, chapter 2), but also implies that P in natural apatite is of limited bioavailability. Indeed, under most conditions it only very slowly dissolves so that plants can use it. Mined apatite contained in crustal rocks must therefore be physically and chemically treated to increase both its concentration and solubility. These upgrading treatments—involved in the preparation of various forms of commercial fertilizer—are commonly referred to as "beneficiation." Initial physical treatment can produce a ground natural apatite concentrate called rock phosphate that, if chemically untreated, can be used in organic agriculture. Rock phosphate is cheaper to transport than raw phosphate-bearing rock; transport of fertilizer is, of course, cheaper still.

Exploration and Discovery

Miners generally explore for rocks called **ore*** with the highest possible concentration of the ore mineral (in this case, apatite). Although rocks consisting of virtually pure apatite occur, they are relatively rare. **Phosphate rock,* PR**, with a weight concentration of P_2O_5 greater than about 10–20 percent, is relatively common and, depending on mining costs and risks, may be considered mineable ore. In comparison, pure apatite has a weight concentration of P_2O_5 of about 42 percent (Manning 1995) or 18.5 percent as P. Lower P_2O_5 contents would qualify a rock simply as "phosphatic rock" (e.g., Chandler and Christie 1995).

Mineable ore grades typically occur in only two rock types: igneous plutonic (rocks crystallized slowly from a melt deep in the Earth), and marine sedimentary (rocks crystallized from seawater and its organisms close to the Earth's surface). Of these two, the marine sedimentary rocks called **phosphorites*** account for the vast majority of current production and reserves, and can be expected to become even more dominant in the future, as igneous deposits are depleted. Large similar deposits have been identified offshore on continental platforms, but accessing them will require technology breakthroughs. Thus, the exploration frontier for P resources seems to lie mostly in technological innovations for accessing known but currently unprofitable deposits rather in the discovery of new deposits.

Geology: Igneous Phosphate Rock and Sedimentary Phosphorite

Igneous rocks sufficiently enriched in apatite to be mined (i.e., phosphate rock) form as the result of two successive processes. First, a very small proportion of the Earth's deep solid mantle must melt, preferentially concentrating P and F in the resulting partial melt. Second, the partial melt must rise very slowly, allowing high-melting-point ("refractory") minerals to crystallize first, thus further concentrating P and F in the melt (a process called fractionation or fractional crystallization). When the residual melt, now of unusual (P-rich) composition, finally crystallizes completely, a large fraction of the crystals formed last will consist of apatite. This mechanism of formation may occur anywhere in the world, and thus igneous rocks enriched in P can be found on most continents. The resulting deposits commonly are elongated vertically as tabular dikes or cylindrical pipes extending deep underground. This vertical geometry renders mining and exploration of these deposits comparatively more difficult than for horizontally layered sedimentary rocks. For this reason, igneous phosphate deposits account for less than 10 percent of world P production (Knudsen and Gunter 2002). Currently, igneous phosphate ores are produced mainly in Russia, the Republic of South Africa, Brazil, Finland, and Zimbabwe (van Kauwenbergh, 2010).

Deposits of sedimentary marine phosphate rock (phosphorite) form under relatively specific environmental conditions and as a result their distribution tends to be geographically restricted. Indeed, most deposits apparently formed within about 40 degrees of the equator, in shallow waters, on stable continental shelves at times when global sea levels were higher than at present. Deposition typically occurred in a two-step process involving the upwelling of large amounts of cold, P-rich deep water into a semi-restricted shallow marine basin in a warm desert region with clear quiet waters that promoted extremely active organic productivity. In such basins, P was biologically taken up and deposited at the bottom with little sand or mud influx (owing to the arid climate) in anoxic (oxygen-free) conditions. This anoxic basin environment favored preservation plus extended **diagenesis*** (post-depositional transformation via dissolution and reprecipitation) and also allowed physical reworking and winnowing of the organic apatite. Currently, relatively low sea levels allow the resulting high-grade marine deposits to be mined cheaply, largely via surface mining. Because of how they were formed, sedimentary marine phosphate deposits (phosphorites) tend to be generally flat-lying, shallow-dipping or horizontal, as well as laterally extensive, and relatively uniform in grade (i.e., P content; see below). Their phosphate content can be comparable to that of the richest igneous deposits (as high as 35 percent P_2O_5), although most are somewhat lower. These features make them relatively easy to discover, drill out, and mine, and it's not surprising that they currently dominate world production, reserves, and resources. However, as mentioned, their distribution is much more geographically restricted than for igneous deposits (e.g., to areas of coastal upwelling that supported high biological productivity in basins adjacent to warm deserts). Phosphorites are the typical Moroccan ores, as well as accounting for most mines elsewhere (as mentioned). Formation of Florida ores apparently was similar, but involved extra surficial reworking (winnowing and transport). The next section explains why only some phosphate rock deposits are considered as valuable sources of P.

The Resource Pyramid: Reserves and Reserve Base

Now we begin to move deeper into the "sustainability circle," moving beyond geological domains to the worlds of technology and economics. Phosphate rock resembles other mined commodities and petroleum, in that its abundance and profitability for extraction can be conceptually plotted on what is commonly called a **resource pyramid*** (or triangle: Figure 3.2). On this conceptual plot (attributed to petroleum geologist J. K. Gray; McCabe 1998, see also Cobb 2009), the horizontal axis (width or area of the pyramid) represents the geological amount present (discussed above), and the vertical axis represents its economic viability (profitability). This economic dimension is a function of a large number of factors (e.g., Brown 2002), such as the ore grade (percent P_2O_5); P selling price (largely a function of demand but also regional cost structures and available production technologies); physical and chemical nature of the ore; depth (directly related to mining cost; e.g., whether a deposit can be mined via highly mechanized strip mining or more costly underground methods); distance from transportation and processing facilities, as well as from agricultural markets; governmental factors (e.g., tax structure, environmental regulations, political stability), plus various other factors, including the availability of financing, miners, energy, and water. Markets for by-products (e.g., U for energy and F for treating municipal water supplies) may also affect profitability.

Any rock that can be mined at a profit is termed ore. The amount of ore that can be profitably and legally mined at any given time via existing technologies is a mine's **ore reserve*** or, more simply, **reserve.*** It is essential to understand from this definition that reserves are therefore highly dynamic and subject to change given price shifts, technological innovations, and changes in the political climate. As a result,

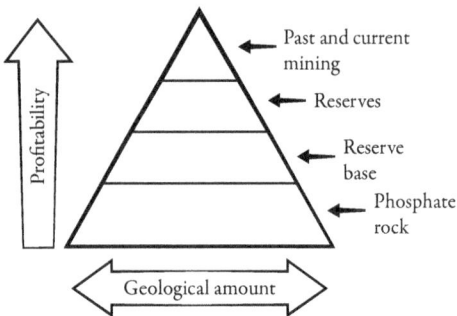

FIGURE 3.2 Schematic representation of the conceptual resource pyramid (profitability vertical, geological amount horizontal), showing that the tonnage of the resource increases as profitability (or grade) decreases. The top cutoff separates out those deposits that have already been mined or are being mined today (only the most profitable or "cream of the crop") from those rocks that could be mined at a profit in the future (reserves), at today's prices and costs. The second cutoff separates profitable reserves from those deposits of marginal grade or profitability (reserve base) that might be mined in the future, should prices increase or technology improve sufficiently. The third cutoff separates this potentially mineable rock (reserve base) from common phosphatic rock that is unlikely to be mined for fertilizer under any likely scenario.

the tonnage of reserves is economically and legally determined and is not solely based on fixed geologic concentrations. Thus, a *reserve* is a "moving target" in that it is constantly changing as the demand for phosphorus and other factors change its profitability and as detailed exploration continues. In contrast, the amount of P in the ground (the *resource* tonnage) is permanently fixed by geological factors, at least over human lifetimes. It is therefore a serious mistake to confuse constantly changing reserve numbers with the ultimate size of the geological resource (e.g., Wellmer 2008), as is commonly done. Furthermore, accurate determination of reserves requires extensive and expensive drilling and assaying on the part of the mining company; the exact amount of exploration drilling required to delineate reserves can be determined by regulations that differ from country to country. The costs of such exploration mean that companies rarely delineate more reserves than are required to raise money via stock sales—typically enough reserves for a few decades of mining. Thus these figures rarely represent the total production potential of the deposit, a phenomenon resulting in reserve growth over time, as additional exploration is paid for out of current profits.

Another moving target is the so-called **reserve base*** (called by some the "resource base"), typically a much larger figure than reserves. As defined by the U.S. Geological Survey (2009), these represent phosphate rocks that are known to exist, and that meet sufficient profitability criteria (grade, depth, continuity, etc.) that they are likely to be mined sometime in the future, even though they might not be profitable today and might not yet have been explored in detail. Commonly an arbitrary grade cutoff (such as 10 percent or 20 percent P_2O_5) or an arbitrary mining cost is used as a limit on what to include in the reserve base. For P, some authors may include extensive shallow submarine deposits as part of the reserve base, even though the technology to mine them does not yet exist; others may exclude them. In any case, reasonable estimates of the reserve base are far easier for bedded phosphorites, owing to their lateral continuity, than for igneous deposits of phosphate rock.

Although based on a U.S. Geological Survey publication that is now badly outdated, Figures 3.3 and 3.4 give some indication of the actual nature of the resource pyramid for P (Mosier 1986). These respectively depict the cumulative distribution of deposit grades and of deposit sizes for the so-called upwelling-type phosphorite deposits. Figure 3.3, not surprisingly, shows that the largest deposits are relatively rare, although they account for the majority of the tonnage. These features are typical for many ore deposit types. Figure 3.4 shows that 60 percent of known deposits at the time were relatively high grade (more than 25 percent). The sharp decline in abundance from grades of 25 percent and above to grades of 20 percent and below may hint at a relative lack of interest in exploring for lower-grade, higher-cost deposits at the time of the study (1986).

This has changed in recent years, as phosphate rock prices have increased and announced reserves have grown. In fact, the cutoff mining cost criterion for classification of rocks as reserves has more than doubled, to $80–100/tonne (van Kauwenbergh 2010). There have been recent announcements that Algeria, Brazil,

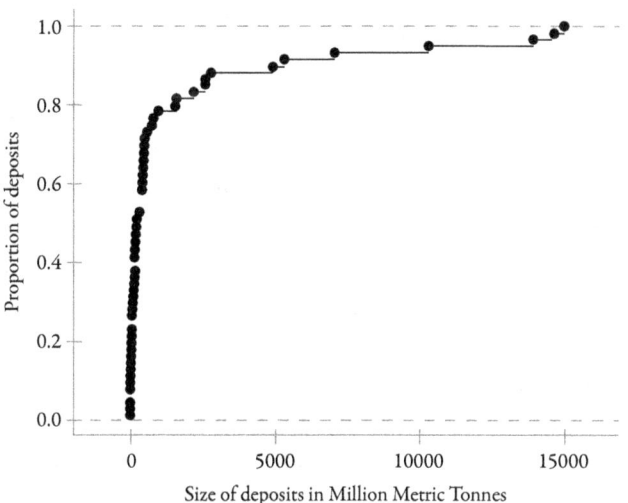

FIGURE 3.3 The cumulative global distribution of phosphorite deposit sizes by proportion. Small (less than 1000 million metric tonnes, and thus costly) deposits accounted for more than 80 percent of known deposits in 1986. In contrast, only a few of the largest deposits account for the majority of the tonnage.

Source: Mosier (1986).

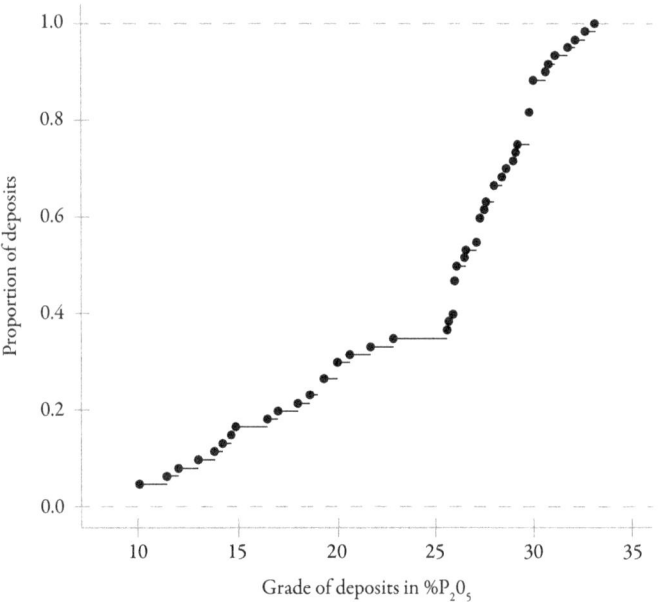

FIGURE 3.4 The cumulative global distribution of deposit grade by proportion. A majority of the deposits known in 1986 contained greater than 25 percent P_2O_5.
Source: Mosier (1986).

China, Israel, Jordan, Syria, and Tunisia are expected to expand existing operations, and new mines are being proposed for development in Australia, Kazakhstan, Namibia, Russia, and Canada (Gurr 2011). These developments are consistent with the notion that increases in price cause the expansion of reserves. In 2010, a new mine opened in Saudi Arabia (Gurr 2011) with a reserve base estimated up to 3000 Mt averaging 18–20 percent P_2O_5 (Zhang et al. 2006). Another one in Peru involves one of the 10 largest phosphate deposits in the world, with reserves above 800 Mt. This was untapped earlier owing to the lack of infrastructure in the remote desert area where it occurs (Zhang et al. 2006). All of these observations are consistent with the concept of a resource pyramid for P, as depicted in Figure 3.2.

During the twentieth century, even lower-grade ores could be mined at a profit, owing to technological advances in mining, as well as to price increases (Wellmer 2008). Most such breakthroughs involved various forms of large-scale mechanization and automation, replacing human labor. These innovations thus depended on abundant and cheap fossil fuels, mainly petroleum. It remains to be seen whether similar technological breakthroughs involving, e.g., nanotechnology or biotechnology, can occur during this century, when energy (and thus mining and farming) costs are expected to increase precipitously. If not, abundant phosphate rock may still remain to be mined, but farmers may no longer be able to afford to buy the fertilizer if food prices remain low. The implication is that large segments situated in the lower section of the P resource pyramid may never be mined.

Having reviewed the meaning of the often misinterpreted terms reserve, reserve base, and resource and having shown how the concept of the resource pyramid can help us understand the evolution of the reserves and reserve base over time, we now turn to current estimates of these quantities.

Reserve and Resource Estimates

The reference used by most scientists, analysts, and journalists for estimates of phosphate rock (PR) reserves and production by country is the annual USGS Mineral Commodity Summary (e.g., Jasinski 2011). These estimates are based on figures given by the geological agencies of phosphate-producing countries, themselves often reporting back reserve estimates from the individual mine. By and large, from 1996 to 2010, the reserves reported by USGS covered 10 countries and hovered between 11,000 million tonnes (Mt) and 18,000 Mt of phosphate rock. Throughout this period Morocco was listed as having the largest reserves (5700 Mt, followed by China and the United States). Most of the fluctuations in reserve estimates were due to changes in the reserve estimates from China. The reserve base was estimated at 147,000 Mt, but because the grade and economic potential are so uncertain, reserve base estimates are no longer reported by the USGS (van Kauwenbergh 2010).

In 2010, the IFDC conducted a more thorough review of publicly available data on phosphate rock reserves, gathering information from a large variety of sources (van Kauwenbergh 2010). The result of this review was a significant upward revision of the estimated reserve, now placed at 60,000 Mt of **recoverable concentrate**,* implying a reserve to consumption **(R/C) ratio*** of 350–400 years, which led the IFDC to conclude that P should be available far into the future.

The largest difference between the USGS numbers and the IFDC numbers concern the reserves of Morocco, now estimated to be 51,000 Mt, based mainly on a publication of the Office Chérifien des Phosphates (OCP, the national mining company of Morocco), after correction for the amount extracted since the date of publication of that report in 1989. The reasons for the discrepancy between the USGS numbers and the IFDC numbers are that the IFDC uncovered data sources that had previously been ignored, and they systematically estimated the ore-to-concentrate ratio of each mine based on the beneficiation technology in use at that mine. The USGS has accepted the IFDC estimates and included them in their 2010 report on global phosphate rock reserves (Jasinski 2011). Table 3.1 shows a summary of these data, along with computed percentages for each country's contribution to the total.

Table 3.1 shows how unevenly distributed the reserves are around the world. Clearly, Morocco and Western Sahara dominate the reserves. Most of Western Sahara, including its phosphorus mines, is occupied by Morocco. Thus, this one nation controls 77 percent of the world's phosphate rock reserves.

It must be noted that the numbers in Table 3.1 are derived from secondary sources and come with considerable uncertainty. The uncertainty stems from unsystematic referencing of the measurement methods, or even the original source of the measurements, and because owners of many of the resources consider detailed data proprietary.

TABLE 3.1
PHOSPHATE ROCK (PR) PRODUCTION AND RESERVE ESTIMATES (JASINSKI 2011)

Country or region	Production (Mt/yr)	(%)	Reserves (Mt)	(%)
Morocco and Western Sahara	26	15	50,000	77%
China	65	37	3700	5.7%
Algeria	2	1	2200	3.4%
Syria	3	2	1800	2.8%
Jordan	6	3	1500	2.3%
South Africa	2	1	1500	2.3%
United States	26	15	1400	2.2%
Russia	10	6	1300	2.0%
Brazil	6	3	340	0.52%
Israel	3	2	180	0.28%
Senegal	0.65	0.4	180	0.28%
Egypt	5	3	100	0.15%
Tunisia	8	4	100	0.15%
Australia	3	2	82	0.13%
Togo	0.8	0.5	60	0.09%
Canada	0.7	0.4	5	0.01%
All other countries	10	5	620	1.0%
World total PR (rounded)	176		65,000	
World total as P	23.0		8509	

The IFDC study constitutes an essential compilation of industry data on reserves but does not provide a geological analysis of deposits nor of the phosphate rock resource base. For such information, one must turn to Project 156, which was carried out in 1986 by the United Nations Environment Program (Cook 1986). The three-volume report of this project contains detailed descriptions of the stratigraphy of deposits in all countries and remains the most comprehensive assessment of PR resources to date. This project estimated that there are 163,000 Mt of resources, equivalent to a lifetime of over 1000 years. However, the study pointed out that most of these deposits are not exploitable with current technology. For example, two-thirds of the PR is found with levels of carbonate minerals that make phosphoric acid production impractical. (Phosphate is extracted from PR by treating it with sulfuric acid to produce phosphoric acid.) Some of the rest of the resources are unsuitable for use with current technology because of contaminants such as cadmium and uranium. Based on this report, one can estimate exploitable reserves to be one-third of the total resources, or 55,000 Mt. This is comparable to the current USGS estimate of 65,000 Mt.

In summary, the USGS Commodity Summary provides annual compilations of publicly disclosed industry data, which gives an estimate of reserves, a dynamic quantity. The USGS data sources and estimation methods were complemented by the recent and influential IFDC report, which caused an upward revision of estimated reserves to 60,000 Mt of recoverable concentrate. Project 156 is the reference for the detailed geological analysis of deposits worldwide. Interpretation of which deposits may fall under the reserve or resource category depends on current economic and technological variables, and thus has to be updated over time.

Reserves of Phosphorus in Soils

Phosphorus naturally contained in soil can also be considered a significant resource. Indeed, before the introduction of mineral fertilizers, agriculture usually led either to gradual depletion of the total P content of soils or, with careful management, to maintenance of soils' P content. For examples of the former, see Newman (1997) and Buresh et al. (1997). In contrast, during the twentieth century alone, about 550 Mt of P were transferred from P-bearing rocks to soils (Smil 2000), resulting in an enrichment in the total P content of the soils that were most heavily fertilized. As discussed in more detail in later chapters, much of this P tends to accumulate in rather stable forms in soils and later becomes available for plant growth.

Estimates suggest that agricultural soils may hold a total amount of P of the same order of magnitude as the amount of P contained in mineable phosphate rock reserves, although there is considerable uncertainty. A survey from 42 countries of the intertropical zone (Roche 1980) indicated an average total topsoil P content for that climatic region of 400 ppm. Extrapolating this average to the first 50 cm of soil of all the world's croplands (ca. 1500 Mha; according to Ramankutty et al. 2008), we obtain a potential cropland soil P pool of about 4000 Mt. The USGS figure for PR production for 2010 is 176 Mt, or about 23 Mt as P (assuming ore with a 30 percent P_2O_5 content). Thus, this estimate of soil reserves represents about 174 years of production. However, this is certainly an underestimate because 400 ppm is a rather low concentration, indicating depletion (Sanchez 1976).

The most recent data set that includes measures of total soil P content is the USGS Geochemical Landscape Project, an ongoing soil geochemical survey of North America. Two cross-continental transects were performed as part of the pilot phase of the project, yielding 253 data points (Smith et al. 2005). These data show a geometric mean P content in the topsoil of 639 mg P/kg (more or less by a factor of 1.76) for agricultural soils and 410 mg P/kg (more or less by a factor of 2) for other land uses. If we coarsely extrapolate the overall geometric mean of this data set to the whole territory of the United States, we obtain a pool of 3200 Mt of P, which is 38 percent of the global reserve estimate!

Whether or not these long-term soil pools of P can be considered valuable reserves of P for agriculture depends on whether the P contained therein can transfer to the plant-available pool rapidly enough to support growth. The rate at which P is transferred from long-term soil pools to the plant-available pool depends on the bonding strength of the P-containing chemical species present in the soil and many other characteristics of the soil–plant system explained in detail in chapter 2. For example, in highly weathered soils containing large amounts of metal oxides and clays, the predominant reaction is often adsorption of P to these oxides and clays. These typically are strong bonds. As a result, long-term agricultural experiments in the tropics often show nearly immediate declines in yield in the absence of P fertilizer inputs even when the total P level in the soil is high (e.g., Beck and Sanchez 1996). However, once all the strong adsorption sites have been filled (saturated), pools of lower bonding strength can form. Thus, in highly weathered soils, the pool of P must become relatively large before the soil reserves are able to sustain ongoing plant productivity (Dumas et al. 2011). In contrast, on mid-latitudes/northern soils that are not thoroughly weathered and that have a relatively high total P content (>800 ppm), yields can be maintained for at least a decade in the absence of P fertilizer inputs (e.g., Gallet 2002 or Otabbong et al. 1997).

P reserves in soils are valuable because they could allow agroecosystems to persist despite low or discontinued mineral P input from mined ores, as long as this soil reserve is replenished through recycling of organic matter. This implies that our use of P fertilizers is in many cases quite inefficient because the P cumulatively stored in soils is not fully utilized (i.e., P in fertilizer is applied much faster than it can be utilized, so much of it goes into strongly bonded storage or is lost to the hydrological system). The extent to which the soil resource base can become a usable reserve depends also on the progress of agronomic research. Indeed, as explained in chapter 4, researchers are seeking ways to enhance plants' abilities to uptake P from the large stable soil P reserves (see review by Jansa et al. 2011 on enhancing mycorrhizal activity for P acquisition, and Yang et al. 2007 for work aiming at up-regulating pyrophosphate expression and Crews 2005 for the advantages of perennials over annuals in allowing tighter nutrient cycling). In conclusion, given their size, the soil P pools constitute a non-negligible part of our P resources, although their effective use and maintenance depends on many other factors, which will be explored throughout this volume. In the next section, we will explore the factors that drive global phosphorus production and demand.

PHOSPHORUS PRODUCTION AND DEMAND

Historical Production and Use of Phosphate Rock

Phosphate rock is a primary ore extracted by the mining industry, which is then used as an intermediate product for many other industries. More than three-fourths of this ore goes into production of phosphorus fertilizers. Other uses include cleaners and detergents, mineral supplements to processed food and animal feeds, and toothpastes (Villalba et al. 2008). Also, in some developing regions, phosphate rock is sometimes directly applied to fields without processing.

Many different technologies are used to produce fertilizer from phosphate rock, sometimes differing by the phosphorus content of phosphate rock. The type of fertilizer produced for the global market is hence linked to the availability of different grades of phosphate rock in different regions. For instance, monoammonium phosphate (MAP) and diammonium phosphate (DAP) are two of the most common types of phosphate fertilizer produced globally, and each production process uses phosphate rock of different grades (Leikam 2005). In recent decades, demand for DAP has increased, yet it can only be produced using higher grades of phosphate rock. As higher grades become scarce, fertilizers such as MAP that can be produced with lower-grade PR will become more prominent. However, MAP fertilizers are not as easily transportable and hence make export of raw phosphate rock much less profitable (Leikam 2005), encouraging vertical integration of extraction and fertilizer production locally. The regional availability of different grades of phosphate rock is therefore directly linked to the intermediate demand for certain types of fertilizer.

Table 3.1 shows current production of phosphate rock mining by country (Jasinski 2011). China has recently overtaken the United States as the world's largest producer of phosphate rock, and the top three countries (China, the United States, and Morocco) annually account for 66 percent of the world's phosphate rock production. Morocco is the largest net exporter of phosphate rock, sending the majority to the United States for fertilizer production (IFA 2010). Neither the United States nor China exports phosphate rock but instead send it directly to domestic fertilizer manufacturing facilities. The United States does, however, export manufactured fertilizer. In fact, since 1999 the United States has imported PR from Morocco to maintain its fertilizer production. The United States was responsible for about 25 percent of PR world production during the twentieth century.

If production of phosphate rock, P fertilizers, and its intermediaries is highly geographically patterned, the same can be said of consumption despite the fact that P is a universal input to agricultural production. Indeed, there are large global imbalances in phosphorus fertilizer application per hectare of arable land, reflecting, in large part, differences in the affordability of P fertilizer in different regions. The world average is 100 kg/ha. France uses 240 kg of fertilizers per hectare of wheat versus 25 in Russia, while the United States uses 257 kg/ha of fertilizer for corn against 12 in Tanzania (Office Chérifien des Phosphates, 2009). In sub-Saharan Africa, barely 9 kg of fertilizer are used per hectare! At the global scale, phosphorus application surpluses (calculated as annual P inputs minus annual P outputs) exist on 71 percent of agricultural land, with application deficits occurring on the other 29 percent (MacDonald et al. 2011). Thus, the pricing of P fertilizer seems to lie behind both dimensions of P sustainability (its scarcity and its potential to pollute water bodies): if it's "too cheap," it can be over-applied and produce eutrophication problems; if it's "too expensive," small-scale, low-income, farmers may be too risk-averse or lack access to sufficient credit to invest in this input even if it would be profitable. Understanding the behavioral (e.g., Duflo et al. 2009) and institutional challenges behind regional P scarcity is a major challenge; continuing research efforts in this area will be indispensable in addressing such inequities and establishing the basis for sustainable utilization of soil P resources.

Turning now to a global view, Figure 3.5 shows global PR production from 1900 to 2010, based on USGS data (U.S. Geological Survey, 2010; Jasinski 2011). The fastest decade of increase was during the post-WWII period from 1944 to 1954, when PR production, on average, increased 12.6 percent per year. Another peak decade occurred from 1958 to 1968, when PR production increased an average of 10.8 percent per year. This period corresponds to the Green Revolution, in which increased fertilizer input was one of the essential factors allowing large food yield increases

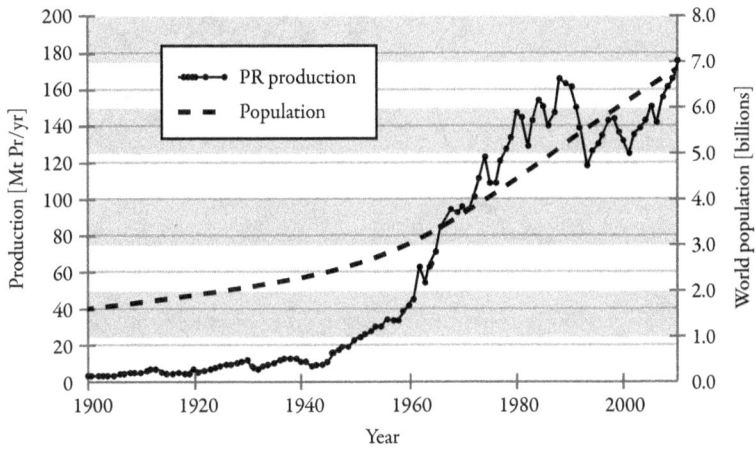

FIGURE 3.5 Global PR production (Jasinski 2011) (solid line joining points) and population (U.S. Census 2011) (dashed line).

in target countries such as India. In contrast, from 1988 to 1993 PR production decreased sharply by 28 percent overall, an average of -6.4 percent per year. Smil (2000) suggests this was in large part due to the collapse of the Soviet Union, but it likely also reflected the recognition in developed countries that yields could be maintained with lower fertilizer application rates. Since then, global production has increased steadily once again. In 2010, total production exceeded the earlier global "peak" of 1987.

The total global PR production of roughly 20 Mt P/yr may be put in perspective by comparing it with estimates of natural flux from weathering of bedrock. Smil (2000) estimated the total flux of phosphorus from natural erosion to be at least 10 Mt P/yr, of which at least 7 Mt P/yr is transported to the oceans by rivers (the balance would accumulate in terrestrial deposits). Thus the P flux from mining (anthropogenic "erosion") would appear to be roughly double the flux from natural erosion.

How is global production/consumption of P changing? Trends in human P consumption may be better understood by separating them into different components, such as population increase, dietary changes affecting per capita demand, and diversion of cropland for biofuel. Figure 3.6 shows per capita PR production (PR/N) since 1940.

Inasmuch as the major factors that contributed to the high PR/N up to 1988 are no longer operational and we are interested in current trends, we focus on the period from 1993 to the present, as shown in Figure 3.7, and ask whether these data can help us assess whether there is a significant increasing trend in PR production. The

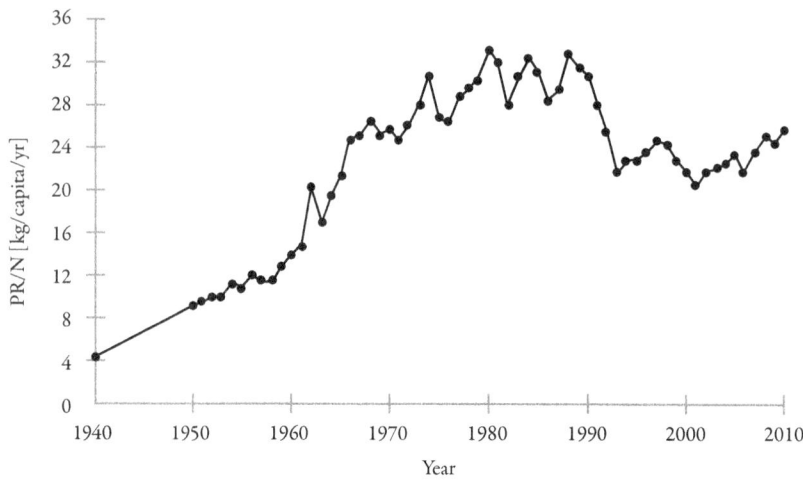

FIGURE 3.6 Global per-capita phosphate rock production.

trend for the entire period from 1993 to 2010 is not significantly different from a zero growth rate. However, the trend for only the last decade, as shown by the dotted line in Figure 3.7, is statistically significant, with an average rate of increase in per-capita PR consumption of 2.5 percent per year.

These data by themselves cannot tell us whether to accept a zero-growth trend since 1993 or the strong growth trend indicated for the period from 2001 to 2010. This choice would require making assumptions grounded in knowledge of the factors affecting the demand for and production of phosphate rock. For example, there are at least two factors supporting the existence of a significant medium-term increase in demand for PR. These are economic growth in developing countries such as China, which should increase the demand for meat products, and the increased use of crops for fuel. The increase in global population plus increased resource demand from China and other countries could make one skeptical of the zero-growth scenario and support choosing the trend based on 2001 to 2010, given no dramatic technological changes in the near future.

Prices of PR are related to these trends in production of PR and demand for goods (such as agricultural products) that require PR as an input. Global prices of phosphate rock vary by region and grade, but generally declined in decades prior to 2007 or so, and were as low as $8 per tonne in the United States in 2000. However, prices exploded in 2007, from $45 per tonne in April 2007 to $430 per tonne in August 2008 and to as high as $500 per tonne in North Africa in 2009 (Van Kauwenbergh 2010). They have declined again since that peak, but now are trending upward again, to $182 per tonne. These extreme price variations have triggered new interest in the availability of

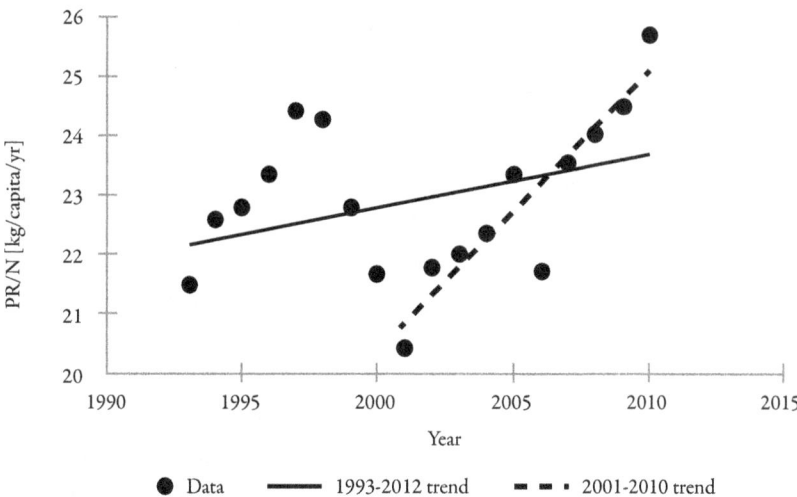

FIGURE 3.7 Global per-capita production of phosphate rock from 1993 to 2010. The 1993–2010 trend line was not significantly different from zero growth for that period.

phosphorus and have revealed the inflexibility of mine production in the short term. Increasing mining and processing costs, due to (for instance) increasing energy costs or declining grades plus increasing global demand for agricultural products have the potential to keep the cost of phosphate rock and fertilizer relatively high unless investments in new technology are able to bring down production costs for the marginal PR producers. However, note that world PR capacity is projected to increase by ~25 percent by 2015 due to the opening and expansion of projects throughout the world (IFA 2011). Similarly, 34 new phosphoric acid production units are opening and about 40 new MAP.

THE FUTURE OF PHOSPHATE ROCK

As described above, the world has witnessed large fluctuations in the production and prices of both phosphate rock and P fertilizers. Particularly violent spikes at the end of the last decade have sparked concerns regarding potential future scarcity of mineral phosphorus. Thus, to conclude this chapter, we will describe several approaches for forecasting future demand and production and discuss the limitations of each. In doing so we wish to emphasize that "P scarcity" should not be seen as a *geological* problem but rather as a *geopolitical* and *economic* question. This is because the key issue is price, not the absolute amount of P in the ground. In addition, while no criterion for what should constitute a sustainable P system can authoritatively be stated without a broader public debate and philosophical investigation of the problem, we will conclude with our own thoughts about what sustainability entails and assess whether expected developments are in accordance with it.

A very simple way to project the future availability of a resource is to consider the ratio of reserves to the consumption rate (R/C). For example, considering the data in Table 3.1, the reserve/consumption ratio would be R/C = (65,000 Mt PR)/(176 Mt PR/yr) = 369 years. However, this index must be used with caution; it has been taken too literally by some. It would only be meaningful if both numerator and denominator were expected to be stable for the computed time period. Of course, reserve estimates depend on the production cost structures associated with different deposits, regions, and institutions. New resource discoveries and changing consumption rates are subject to numerous influences as well. Although considered by some to be a naïve index, the R/C ratio can be used as a benchmark for estimating phosphorus availability. It shows that although phosphorus reserves will not be depleted in the near future, their long-term availability depends either on our ability to increase reserves in the future through new mining efforts, exploration, and technological advancements, or on our ability to at some point transition to an agricultural system that is not as dependent upon extraction of phosphate rock.

Another approach involves fitting a standard curve to historic production data based on the assumption that resource exploitation tends to increase initially, reaches a maximum as limiting conditions set in, and then declines steadily. This produces a "bell-shaped curve." The method was pioneered by M. King Hubbert to predict oil production, and accordingly is called the Hubbert curve method. Cordell et al. (2009) used the Hubbert curve to extrapolate PR production into the future. The area under the Hubbert curve is the stock, and the slope of the curve is the rate of extraction at any given time. Fitting this model to pre-2010 USGS estimates of reserves, Cordell et al. (2009) predicted that a peak P would be reached in 2033. Fitting it to the new data from the IFDC pushed the peak further into the future, but only by just over a century (to 2145), not several centuries, bringing Cordell et al. (2009) and Vaccari (2009) to the conclusion that emerging P scarcity should still be a concern.

The Hubbert curve method is subject to some of the same limitations as the R/C ratio. It assumes that the production curve will be roughly symmetrical in time. The most common utilization of the Hubbert curve has been for oil, for which it is still a debated technique. Vaccari and Strigul (2011) have shown that Hubbert curve fits are highly sensitive to the stage of production. In the example of U.S. production, for example, reliable fits could be obtained only after the peak had already occurred.

Ideally, forecasts should be made using dynamic models, taking into account actual mechanisms of future change. For example, one could consider population

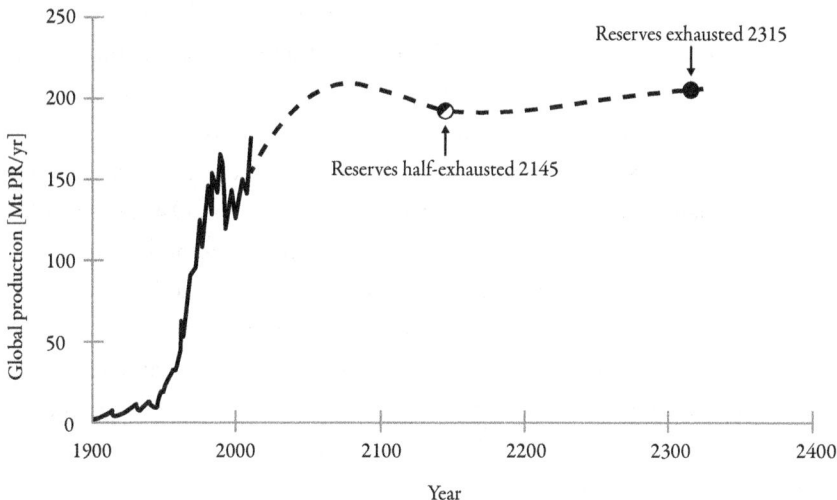

FIGURE 3.8 Future phosphate rock production based on population projections and assuming current per-capita consumption.

increase, which will have a direct effect on demand for food and hence an indirect effect on demand for fertilizers and phosphate rock. Demographic and economic data could be used to estimate both the future population and per-capita components of demand. Of course, this method is subject to its own limitations, especially because those causal factors have their own sources of uncertainty.

As an example, Figure 3.8 shows a forecast of future PR production that uses the UN estimate for population growth but assumes that per-capita consumption remains at current levels. This shows that the point at which half of the initial reserves present in 1900 would be exhausted would occur in the year 2145. This point would correspond to the peak value for the Hubbert model, and agrees with it numerically. Therefore, both these approaches set a benchmark of 100–150 years of P resources before they may become physically and technically constrained. From this benchmark, one can perform further scenarios, estimating changing future consumption or reserves using boundaries suggested by experts on future physical, technical, and economic potentials.

Although all of these forecasting methods are fraught with difficulties, they represent the best efforts of which we are currently capable. An advantage of models is that the assumptions are laid bare for discussion. Those who disagree can propose other assumptions or other models, advancing the overall discussion of P sustainability and scenario modeling in general.

What do these highly imperfect forecasts tell us about the sustainability of P-use? Ideally, sustainability would be achieved by taking an infinite planning horizon.

However, searching for a system that can be maintained indefinitely is illusory because the world is in constant flux: all major factors affecting the future of the demand and supply of P are irreducibly uncertain. What seems more useful as a criterion that we *can* handle given our limited view of the future is the criterion that the agricultural system remain as *adaptive as possible,* i.e., that we strive to keep more, rather than fewer, options open to reduce the costs of adapting to unforeseen circumstances. The current system brings about the waste of fertilizers due to over-application and erosion in many northern agro-systems and degradation of soils due to under-application of fertilizers in many southern agro-systems. These two trends make agricultural systems less adaptive. Indeed, wasteful consumption of high-grade and easily accessible ore will raise the material and organizational requirements for ensuring the same level of accessibility in the future, whereas degradation of soils reduces the area of arable land.

In summary, phosphate mining, especially surface mining, is environmentally and energetically costly, and, as we explained in this chapter, many factors can lead these costs to increase. Of paramount concern is the effect of regional cost structures on the entire fertilizer production and delivery process, including changes in the price of inputs (such as energy) and the increase in demand. Although global markets have the potential to adapt to these changes, such institutions are limited by a number of physical, social, and environmental constraints, and we cannot be certain that price signals will create a steady transition to a sustainable phosphorus system that minimizes human suffering along the way. Institutional dimensions constraining societal response to the myriad dimensions of P sustainability are discussed in depth in chapter 8. The sooner we can understand these constraints and adapt our current institutions accordingly, the less likely it is that we will face the prospect of a phosphorus crisis, should circumstances force it upon us.

BOX 3.2
CHAPTER 3 SUMMARY

- Phosphorus is found in almost all rocks and soils on earth, but a few large deposits dominate economic phosphate rock production.
- Phosphorus reserves are economically determined and change with time; resources are geologically determined and do not change with time. The distinction is underappreciated.
- Phosphorus resources in soils appear to be of the same magnitude as reserves in rocks.

BOX 3.2 (*Continued*)

- The most conservative way to forecast depletion of economic phosphorus reserves is by comparing them to the current consumption rate; however, this should be considered only a benchmark for potential scarcity, not a predictor of future scarcity.
- Empirical methods such as Hubbert's curve may also be useful as a benchmark, but their implicit assumptions are not always valid.
- The world's economic reserves are estimated to be several hundred times the current annual consumption, but estimates based on the median projection of population changes show that half of known reserves could be depleted in a little more than a century.
- As higher-grade resources become depleted, lower-grade resources can be substituted; however, these may become costly and technologically constrained, rendering virgin production unsustainable.
- The consumption trend over the past decade has been an increase of 3.6 percent per year, of which 2.5 percent per year is the per capita increase. Consumption is highly heterogeneous, owing to differences in affordability of P fertilizers: many regions consume more than the environmentally desirable level, while many others consume less than sustained soil fertility would require.
- Therefore, absolute shortages of phosphorus are not an issue in the medium term; the relative cost structures of production are.

REFERENCES

Beck, M. A., and P. A. Sanchez. 1996. Soil phosphorus movement and budget after 13 years of fertilized cultivation in the Amazon basin. *Plant and Soil* 184: 23–31.

Brown, W. E. 2002. Volume IV: "Sociocultural and Institutional Drivers and Constraints to Mineral Supply." In *The Meaning of Scarcity in the 21st Century: Drivers and Constraints to the Supply of Minerals Using Regional, National and Global Perspectives*, U.S. Geological Survey Open-File Report 02-333.

Buresh, R. J., P. A. Sanchez, F. Calhoun. 1997. Replenishing Soil Fertility in Africa. *Soil Science Society of America*, Special Publication No. 51.

Burke, E. A. J. 2008. The use of suffixes in mineral names. *Elements* 4(2): 96.

Chandler F. W. and R. L. Christie. 1995. Stratiform phosphate, in *Geology of Canadian Mineral Deposit Types*. In Geological Survey of Canada, Geology of Canada, No. 8, edited by O. R. Ekstrand, W. D. Sinclair, and R. I. Thorpe, 33–40.

Cobb K. D. 2009. http://scitizen.com/future-energies/energy-the-achilles-heel-of-the-resource-pyramid_a-14-2760.htm.

Cook, P. J., and J. H. Shergold, (eds.). 1986. *Phosphate Deposits of the World*, Cambridge University Press, New York.

Cordell, D., J. O. Drangert, and S. White. 2009. The story of phosphorus: Global food security and food for thought. *Global Environmental Change* 19(2): 292–305.

Crews, T. E. 2005. Perennial crops and endogenous nutrient supplies, Renewable Agriculture and Food Systems 20(1): 25–37.

Duflo, E., M. Kremer, and J. Robinson. 2009. Nudging farmers to use fertilizer: theory and experimental evidence from Kenya. Bread working paper No. 233.

Dumas, M., E. Frossard, and R. W. Scholz. 2011. Modeling biogeochemical processes of phosphorus for global food supply. *Chemosphere* 84:798–805.

Gallet, A. 2002. *Phosphorus availability and crop production in seven Swiss field experiments*. PhD thesis, ETH.

Gurr, T. M. 2011. Phosphate rock, *Mining Engineering* 63(6):88–91.

Huminicki, D. M. C., and F. C. Hawthorne. 2002. The crystal chemistry of the phosphate minerals. In *Phosphates: Geochemical, Geobiological, and Materials Importance,* edited by M. J. Kohn, J. Rakovan, and J. M. Hughes. *Reviews in Mineralogy and Geochemistry* 48:123–253.

IFA. 2010. Global fertilizer trade flow map. http://www.fertilizer.org/ifa/HomePage/FERTILIZERS-THE-INDUSTRY/Global-fertilizer-trade-flow-map

IFA. 2011. Fertilizer Outlook, 2011–2015.

Jansa, J., R. Finlay, H. Wallander, F. A. Smith, and S. E. Smith. 2011. Role of Mycorrhizal Symbioses in Phosphorus Cycling. In: *Phosphorus in action—Biological processes in soil phosphorus cycling,* edited by E. K. Bünemann, A. Oberson, E. Frossard. Springer-Verlag, Berlin Heidelberg, Germany, 137–168.

Jasinski, S. M., 2011. Phosphate Rock [Advance Release], *USGS Minerals Yearbook 2010,* 56.1–56.10.

Knudsen, A. C., and M. E. Gunter. 2002. Sedimentary phosphorites—An example: Phosphoria Formation, Southeastern Idaho, U.S.A. In *Phosphates: Geochemical, Geobiological, and Materials Importance: Reviews in Mineralogy and Geochemistry* edited by M. J. Kohn, J. Rakovan, and J. M. Hughes, 48:363–389.

Leikam, D. F., and F. P. Achorn. 2005. Phosphate fertilizers: production, characteristics, and technologies. *Phosphorus: Agriculture and the Environment,* edited by J. T. Sims and A. N. Sharpley, Agronomy Monograph No. 46, 23–50.

MacDonald, G. K., E. M. Bennett, P. A. Potter, and N. Ramankutty. 2011. Agronomic phosphorus imbalances across the world's croplands. *Proceedings of the National Academy of Sciences*, 108, 3086–3091. doi:10.1073/pnas.1010808108.

McCabe P. J. 1998. Energy resources—Cornucopia or empty barrel. *American Association of Petroleum Geologists Bulletin* 82:2110–2134.

Manning, D. A. C. 1995. *Introduction to Industrial Minerals*. Chapman and Hall, London.

Manning, D. A. C. 2008. Phosphate minerals, environmental pollution, and sustainable agriculture. *Elements* 4(2):105–108.

Mosier, D. L. 1986. Grade and tonnage model of upwelling phosphate deposits. In *Mineral Deposit Models: U.S. Geological Survey Bulletin,* edited by D. P. Cox and D. A. Singer, 234–236.

Newman, E. I. 1997. Phosphorus balance of contrasting farming systems, past and present: can food production be sustainable? *The Journal of Applied Ecology* 34(6):1334–1347.

Office Chérifien des Phosphates, 2009. Annual Report.

Otabbong, E., J. Persson, O. Iakimenko, and L. Sadovnikova. 1997. The Ultuna long-term soil organic matter experiment. *Plant and Soil* 195:17–23.

Pan, Y. and M. E. Fleet. 2002. Compositions of the apatite-group minerals: Substitution mechanisms and controlling factors. In *Phosphates: Geochemical, Geobiological, and Materials*

Importance, edited by M. J. Kohn, J. Rakovan, and J. M. Hughes, *Reviews in Mineralogy and Geochemistry*, v. 48, 13–49.

Pasteris, J. D., B. Wopenka, and E. Valsami-Jones, 2008. Bone and tooth mineralization: Why apatite? *Elements* 4 (2): 97–104.

Ramankutty, N., A. T. Evan, C. Monfreda, and J. A. Foley. 2008. Farming the planet: 1. Geographic distribution of global agricultural lands in the year 2000. *Global Biogeochemical Cycles*, 22, GB 1003, doi 10.1029/2007GB002952.

Roche, P., L. Griere, D. Babre, H. Calba, and P. Fallavier. 1980. *Le phosphore dans les sols intertropicaux: appreciation des niveaux de carence et des besoins en phosphore*. Technical report, Institut Mondial du Phosphate, 1980.

Sanchez, P. A. 1976. *Properties and management of soils in the tropics*. John Wiley and Sons, New York.

Smil, V. 2000. Phosphorus in the environment: Natural flows and human interferences. *Annual Review of Environment and Resources* 25: 53–88.

Smith, D. B., W. F. Cannon, L. G. Woodruff, R. G. Garrett, R. Klassen, J. E. Kilburn, J. D. Horton, H. D. King, M. B. Goldhaber, and J. M. Morrison. 2005. *Major- and Trace-Element Concentrations in Soils from Two Continental-Scale Transects of the United States and Canada*, USGS Open-File Report 2005-1253.

US Geological Survey. 2010. *Mineral Commodities Summaries*, Washington, DC.

Vaccari, D. A. (2009). Phosphorus: A Looming Crisis. *Scientific American* 300 (6): 54–59.

Vaccari, D. A. and N. Strigul. 2011. Extrapolating phosphorus production to estimate resource reserves. *Chemosphere* 84: 792–797.

Van Kauwenbergh, S. V. 2010. World Phosphate Rock Reserves and Resources. *Technical Bulletin IFDC T-75*. International Fertilizer Development Center, Muscle Shoals, AL.

Villalba, G., Y. Liu, H. Schroder, and R. U. Ayres. 2008. Global phosphorus flows in the industrial economy from a production perspective. *Journal of Industrial Ecology* 12(4): 557–569.

Wellmer, F. W. 2008. Reserves and resources of the geosphere, terms so often misunderstood. Is the life index of reserves of natural resources a guide to the future? *Z. dt. Ges. Geowiss* 159(4): 575–590.

Zhang, P., R. Wiege, and H. El-Shall. 2006. Phosphate rock. In *Industrial Minerals and Rocks*, 7th ed., edited by J. E. Kogel, N. C. Trivedi, J. M. Barker, and S. T. Krukowski, 703–722, Society for Mining, Metallurgy, and Exploration, Littleton, CO.

4

Sustainable P in Agriculture: Food and Fuel

P IS FOR PRODUCTIVITY

Val H. Smith, Cecil W. Forsberg, Roberto A. Gaxiola, Thomas W. Crawford, Jr., Andrew R. Sharpley, Laura Schreeg, Ben Chaffin

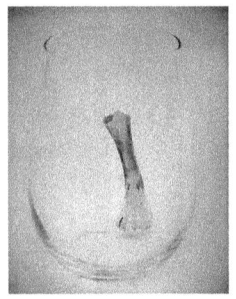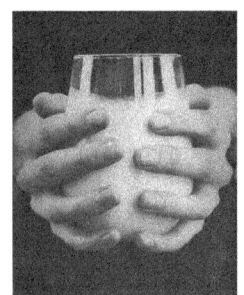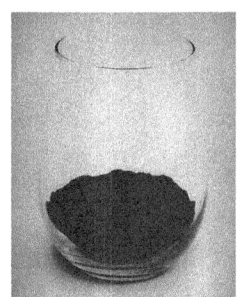

Joshua White; Got Bones? Got Ca? Got P?; Cyanotype, 20" x 40", 2011

Joshua White
Got Bones? Got Ca? Got P?

Scientific Collaborator:
Genevieve Metson, MS student, School of Sustainability, Arizona State University

Description of Artwork:

Phosphorus is an essential element for life, and is a major component of bones. However, when we think of bones we most often think of the element calcium, not phosphorus. This piece is a play on the Got Milk? commercials, where we think of the calcium in milk as important for healthy bones. Here we add the question Got Phosphorus? Perhaps on the scale of one bone, one person, we do have phosphorus. But global mineral phosphorus reserves, the main source of phosphorus for fertilizer in agricultural production, are being depleted. And as a society we must ask ourselves where phosphorus for food production, and ultimately our nourishment and the growth of our bones, will come from. How can we find sustainable ways to recycle and manage P resources for the future?

The left-hand image represents the question Got Bones? Yes. The middle image represents the question Got Calcium? Yes. The right-hand image represents the question Got Phosphorus? We don't know. The glass with milk in the middle representing calcium is full, but the glass on the right with phosphorus is much lower, representing phosphorus depletion. In addition to the content of the pictures, they are cyanotypes. This process results in an image made with iron, which is an element that binds strongly to phosphorus. This finish also represents how global phosphorus availability is decreasing by invoking the chemical bonding of phosphate and iron, making phosphorus unavailable.

About the Artist:

Joshua White received a BFA in Photography from Northern Kentucky University. His work has been exhibited regionally in Ohio and Kentucky, as well as in California and Arizona. As a graduate student at ASU majoring in photography, his work focuses on memory and the way photographs function within that context. The photographs he creates are not clean, straightforward images that point and say, "Look here"; rather, they act as catalysts, as stand-ins for what it feels like to remember.

BOX 4.1
CHAPTER 4 OBJECTIVES

- Quantify and examine phosphorus demands associated with food production and with renewable biofuel production.
- Explore innovative solutions for improving phosphorus-use efficiency at the farm scale, in crops, and in livestock.

PHOSPHORUS DEMANDS IN AGRICULTURE AND BIOFUELS

The challenges of human dependence on phosphorus are of fundamental importance because P is required for the production of both food and renewable bioenergy resources (Childers et al. 2011). The primary goal of this chapter is to explore the phosphorus dimensions of food and biofuel production, and to present potential solutions that may help to increase the phosphorus-use efficiency of these two key facets of human activities (Figure 4.1).

Phosphorus Flows in Food Production

Phosphorus inputs to agricultural production systems are derived from three primary sources (Figure 4.2): inorganic fertilizer P, natural soil P, and applied manure P. However, far less than 100 percent of these three sources of P can be taken up via **assimilation*** into crop plant tissues, and as a result the assimilated fraction of fertilizer P, soil P, and manure P can be relatively low. Similarly, the **phosphorus-use efficiency*** by which a plant transforms this assimilated P into new plant tissue can

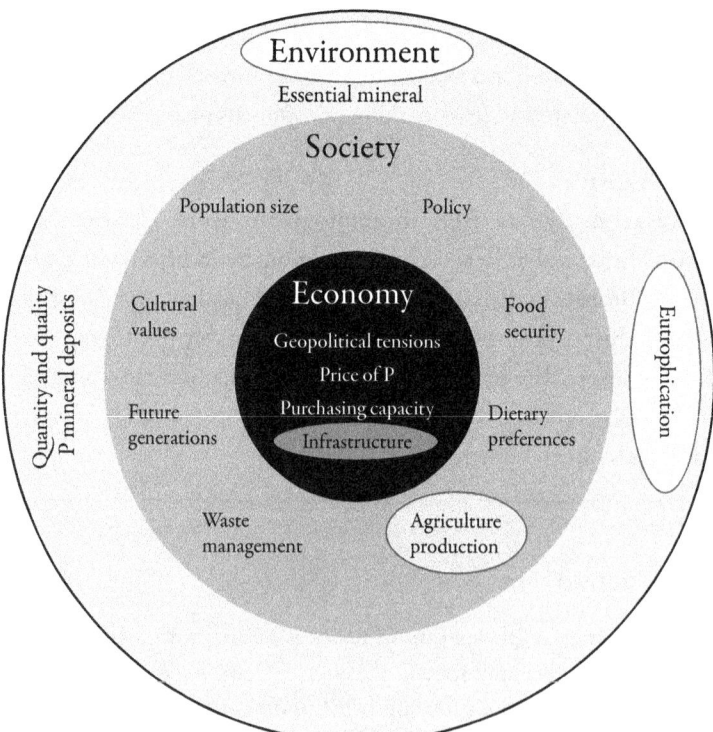

FIGURE 4.1 Agricultural demand for phosphorus works in all three spheres of sustainability thinking. Phosphorus plays a large role in environment (fertilizer use), society (agriculture and diet), and the economy (trade of agricultural goods).

be low and highly variable (PUE, expressed in grams of dry plant weight that can be produced per gram of plant-assimilated P). By analogy, the phosphorus-use efficiency by which grazing livestock can transform ingested P into new animal tissue also can be low and highly variable (PUE, expressed in grams of dry animal weight that can be produced per gram of ingested P).

P incorporated into agricultural production can then leave the production system both as harvested plant and animal materials, and can flow both into and through the agricultural landscape in the form of animal excreta (Figure 4.2). Typically, less than one-third of the P imported as feed into confined animal production systems will be retained within the livestock, with the remaining balance being excreted as urine, as well as contained in manures that are frequently retained for later land application as fertilizers (Patterson et al. 2005; Poulsen 2000; Valk et al. 2000). In addition, phosphorus is exported from the agricultural landscape both in the form of harvested seeds and grains, and as incidental losses in **stover,*** leaves, and straw.

As can be seen in Figure 4.3, more than tenfold variability exists in the phosphorus content of the world's top 15 agricultural crops, and these data have been used to estimate the phosphorus-use efficiency of each of these 15 species; these data are shown in Figure 4.4(a,b). Clearly, the phosphorus-use efficiency of bananas (ca. 1600 g dry wt./g P) is far higher than the least P-efficient food crop (tomatoes: ca. 250 g dry wt./g P) and the least

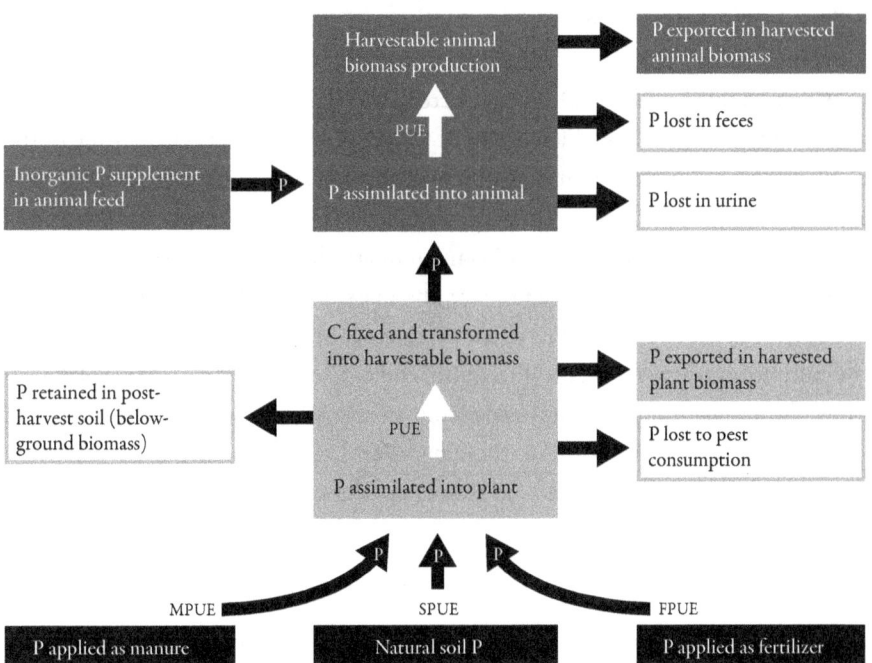

FIGURE 4.2 Phosphorus inputs, phosphorus losses, and phosphorus-use efficiency in crop and livestock production.

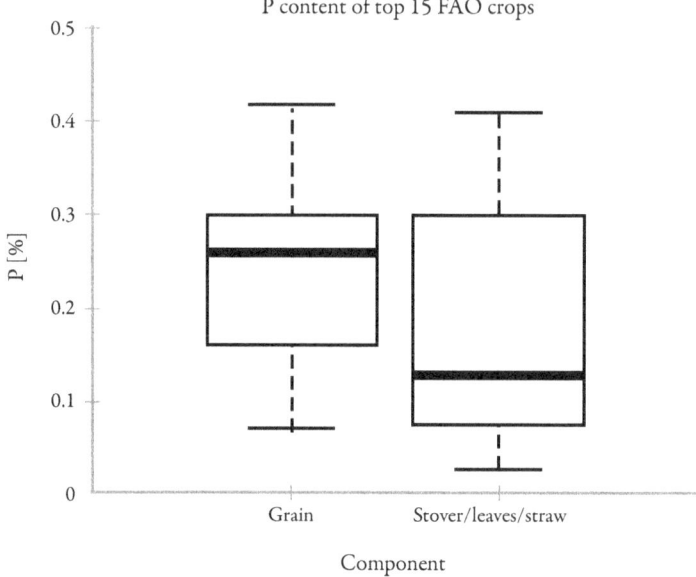

FIGURE 4.3 Phosphorus content of the top 15 FAO crops, expressed as %P in the dry weight of harvested seeds or fruit, and in the associated plant parts that are incidentally removed from the field during the harvesting process.

P-efficient oil crop (camelina: ca. 80 g dry wt./g P). Thus, a given quantity of available P will result in higher yields of the higher PUE crops; similarly, a given quantity of dietary P will result in higher yields of higher PUE livestock.

When coupled with FAO-reported trends in agricultural production, the data shown in Figure 4.4 allow calculations of the total P exported via harvesting of these crops during the past 50 years. This new analysis (Figure 4.5) suggests that total P exports associated with agricultural harvests have increased dramatically during the past five decades, and that they have been dominated by four primary crops: sugarcane, wheat, corn (maize), and rice. However, the mass of phosphorus that is removed in plant harvests represents only 10 to 40 percent of the total P that was applied during the cultivation year as fertilizers (Pierzynski and Logan 1993; Sims and Sharpley 2005). Thus, the annual total phosphorus demand associated with the agricultural production of these 15 crops, expressed as the sum of fertilizer P plus the plant P exported as harvest, can be expected to be 2.5–10 times the values shown in Fig. 4.5.

Phosphorus Implications of Biofuel Production in the United States

Major shifts in crop production and agricultural land use often occur due to external pressures. One of these pressures is a recent renewable-energy-driven shift toward production of biofuels derived from plant feedstocks such as terrestrial crops and

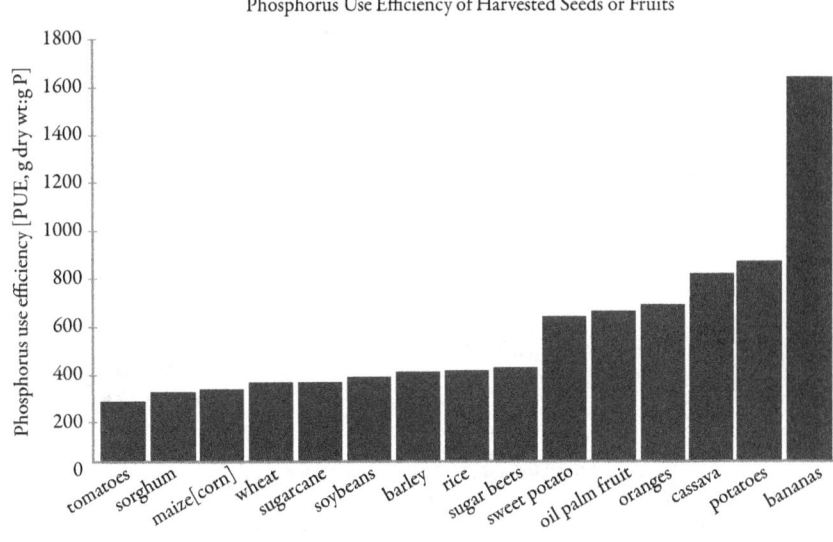
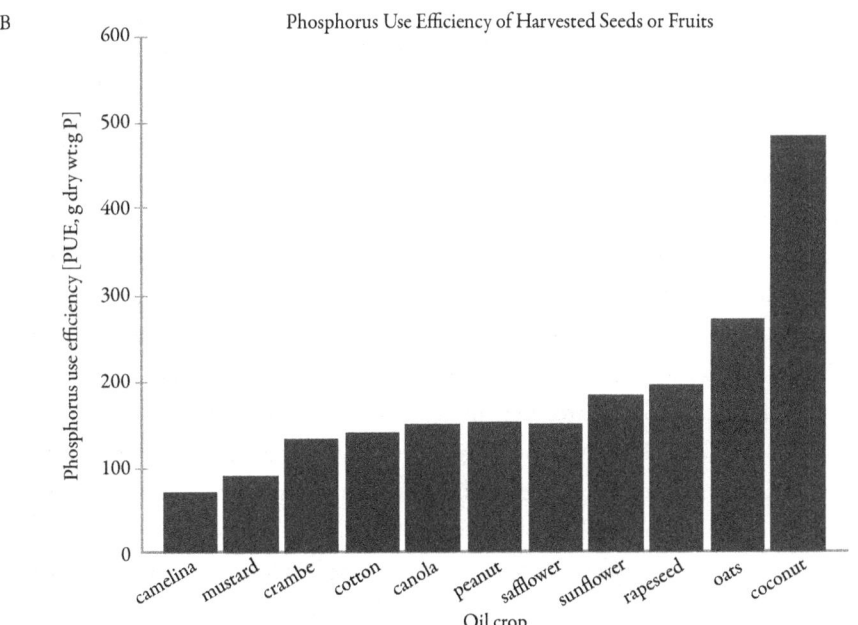

FIGURE 4.4 Phosphorus Use Efficiency (PUE) of major seed and fruit crops, expressed as g dry weight per gram of P in the harvested plant matter. A. Food and bioethanol crops; B. Oil crops.

algae. For example, current biofuels policy in the United States is strongly influenced by the Renewable Fuels Standard (RFS) of the Energy Independence and Security Act of 2007, which mandates a biofuels production target of 36 billion gallons per year (BGY) by 2022. A majority of this target (31 BGY) is to be met

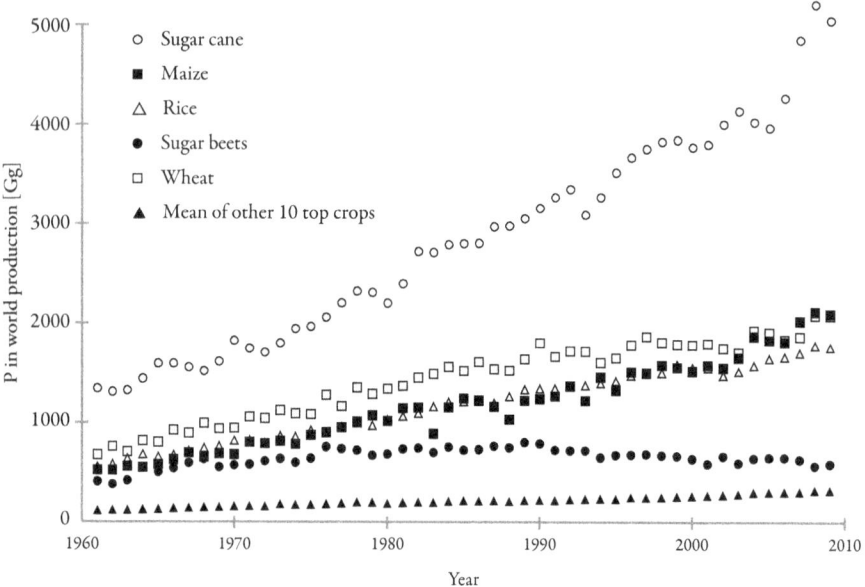

FIGURE 4.5 Time trends in estimated phosphorus demands by the FAO's top 15 food crops.

with ethanol from starch grains and lignocellulosic biomass, and the balance will consist of 1 BGY of biomass-based diesel and 4 BGY of other advanced biofuels (Pate et al. 2011).

Clearly, bioenergy is becoming a socioeconomic necessity, and cellulosic biofuel production thus will almost certainly accelerate in the future. As a result, P demands by crops used for bioethanol production can also be expected to increase. For example, massive increases in the acreage of planted corn occurred in the Chesapeake Bay Watershed and in Mississippi River Basin states between 2005 and 2010 (USDA—National Agricultural Statistics 2010). A majority of this increase in corn acreage resulted either from the conversion of land previously in soybean cultivation and/or pasture, or from land returned to corn production from Conservation Reserve Program (CRP) acreage. Unfortunately, current corn production practices are associated with highly inefficient nutrient use: average losses of P in runoff from corn (2–14 kg P ha^{-1} yr^{-1}) can be very high and tend to be greater than from soybean (2–8 kg P ha^{-1} yr^{-1}; Carpenter et al. 1998; Kimmell et al. 2001; Sharpley and Rekolainen 1997). In contrast, losses of P from the production of perennials and hay crops are generally far less (0.2–2 kg ha^{-1} yr^{-1}) due to decreased runoff volumes and lower crop P requirements, contributing to smaller requirements by these plants for P applications as fertilizer or manure (Sharpley et al. 2001; Smith et al. 1992).

Assuming that fertilizer application rates will rise in the future in response to high corn prices and farmers' desires to obtain optimum yields, agricultural losses of P can be expected to increase significantly relative to current values. Furthermore, water-quality model simulations of converting acreage to cropland from **Conservation Reserve Program*** lands or from perennial grasslands, confirm that delivered P loads can increase by more than twice the percentage of total land area that has been returned to crop production (Mankin et al. 1999, 2003). Assuming that fertilizer application rates per acre remain constant, increasing trends in corn acreage are projected to enhance annual P loads by 0.55 million kg P (220 percent increase) in the Chesapeake Bay Watershed (United States) and by 34 million kg P (200 percent increase) in the Mississippi River Basin (United States); similar P losses can be expected to occur anywhere around the globe where corn production is increased in response to increasing cellulosic biofuel production.

In addition to increasing P demands driven by bioethanol production, rapid growth of grain-based ethanol production has the potential to create further P imbalances through its by-products. **Dry distiller's grain (DDG)*** is a by-product of ethanol production that can be used in animal feed (mainly **ruminants***). Coupled with higher corn prices for feed, the rapidly expanding supply of DDG has resulted in lowered DDG prices, and this supplement is being widely incorporated into dietary rations at substantial rates for beef and dairy cattle. In addition, the use of DDG as a feed ration alternative is likely to increase even further due to its high availability and low cost. However, the P content of DDG (0.8–0.9 percent) is about three times that of corn. Simpson et al. (2008) calculated that at the maximum 20 percent of ration as dry-matter content from DDG, the 1.5 million dairy and 1 million beef cows in the Chesapeake Bay Watershed alone could consume about 5 Mt of DDGs. The authors further projected that replacement of soybean meal, corn, and corn silage with DDGs would equate to about 200,000 kg of additional P in manure. Even with <20 percent DDG supplementation of dairy cow diets, this elevates animal ration P to 0.5 percent P, a value that is higher than the recommended range of 0.33–0.36 percent P (National Research Council 2001), offsetting reductions in dietary P consumption over the past decade that have been gained via feed management. Inclusion of DDG in animal rations at rates such as these thus will increase the P content of manure (Baxter et al. 2003; Maguire et al. 2004; Wu et al. 2001), and if land applied, will also increase the potential for P losses in runoff (Ebeling et al. 2002; Maguire et al. 2007; Sharpley et al. 2005). This trend would likely erode recent progress in managing P in animal feeds, and could make government incentives to improve feed management less attractive to farmers.

The above situation is further complicated by projected future demands for transportation biofuels. For example, a recent analysis of the potential resource

TABLE 4.1

ESTIMATES OF THE TOTAL PHOSPHORUS DEMANDS (IN MILLIONS OF METRIC TONNES OF P PER YEAR) REQUIRED TO PRODUCE FOUR DIFFERENT TARGET LEVELS OF POTENTIAL BIODIESEL PRODUCTION (10, 20, 50, AND 100 BILLION GALLONS PER YEAR) IN FOUR DIFFERENT GEOGRAPHICAL REGIONS OF THE UNITED STATES DEEMED MOST SUITABLE FOR ALGAL BIOFUEL PRODUCTION: SOUTHWEST [SW]; MIDWEST [MW]; SOUTHEAST [SE]; AND NINETEEN LOWER TIER STATES [NLTS]. THE RANGE OF POTENTIAL BIOFUEL PRODUCTION TARGET VALUES FROM 10 TO 50 BILLION GALLONS PER YEAR (BGY) REPRESENTS THE VOLUME OF ALGAL OIL FEEDSTOCK THAT WOULD BE REQUIRED TO MEET CA. 15 PERCENT TO CA. 80 PERCENT OF TOTAL U.S. DIESEL FUEL DEMANDS, WHILE THE LARGEST TARGET (100 BGY) REPRESENTS THE ALGAL OIL PRODUCTION NECESSARY TO EXCEED THE COMBINED U.S. DEMANDS FOR BOTH DIESEL AND AVIATION TRANSPORTATION FUELS (DATA FROM PATE ET AL. 2011).

Geographical region	Target biodiesel production (billion gallons per year)			
	10 BGY	20 BGY	50 BGY	100 BGY
SW, MW, & SE (50% lipid assumed)	0.8 Mt P yr^{-1}	1.7 Mt P yr^{-1}	4.2 Mt P yr^{-1}	5.3 Mt P yr^{-1}
NLTS (20% lipid assumed)	2.1 Mt P yr^{-1}	4.2 Mt P yr^{-1}	10 Mt P yr^{-1}	21 Mt P yr^{-1}

requirements of algal biofuel production in the United States by Pate et al. (2011) suggests that each metric ton (dry weight) of algal biomass produced will require 12 kg of P. For biodiesel feedstocks having a 50 percent oil content, the P required is equivalent to about 0.083 kg P per gallon of algal oil produced; however, if a 20 percent biomass oil content is assumed, these nutrient requirements increase to about 0.21 kg P per gallon of oil produced. The P demands required to support four production levels of algal biodiesel derived from algae grown in four different geographical regions are shown in Table 4.1. We anticipate that the phosphorus demands associated with equivalent levels of biofuel production derived from *terrestrial* oil crops would exceed the values shown for algal biofuels because of significant leakage of applied fertilizer P from the terrestrial landscape.

The implications of the biofuel P demand projections shown in Table 4.1 are profound. The P requirements needed to support the lowest target production level of 10 BGY, assuming an oil content of 50 percent dry algal biomass, represents one-fifth of the total U.S. consumption of P from phosphate rock in 2006. If one assumes a more probable average oil content of 20 percent biomass, the P requirements for algal biofuel production increase to 51 percent of total U.S. P consumption in 2006. Indirect and unintentional impacts on food prices from this level of competition with other uses of P would almost certainly be unsustainable (Pate et al. 2011), unless major efforts are made to recycle the P that is contained in the post-process, de-oiled plant biomass.

INNOVATIVE SOLUTIONS FOR IMPROVING P-USE EFFICIENCY (PUE) IN CROPS AND LIVESTOCK

The data presented in the previous section underscore the pressures that food and energy production systems place on the global human use of P. It is important that we seek to improve the efficiency by which both plants and animals use both current and future supplies of phosphorus. In particular, focused efforts to increase phosphorus-use efficiency by crops and livestock will improve our ability to reduce net agricultural P demands, and therefore will also serve to increase our ability to reuse and recycle P, and to improve long-term global P sustainability by helping to close the human phosphorus cycle (Childers et al. 2011).

Improving Fertilizer Phosphorus-Use Efficiency

Because levels of plant-available soil P are insufficient to support optimum plant growth and crop yield in many agricultural settings, inorganic phosphate typically must be added to most soils, either as fertilizers or as manures (Figure 4.2). However, phosphate is relatively immobile in most soils, and significant amounts of the added fertilizer P can become fixed or unavailable for plant uptake via precipitation with aluminum ions (acidic soils) or via precipitation with calcium and magnesium ions (alkaline soils) (see chapters 2 and 3). Thus, improving **fertilizer phosphorus-use efficiency*** (FPUE, defined here as grams of plant dry weight produced per unit fertilizer P applied) in crop production thus is one important goal for achieving future phosphorus sustainability (Figure 4.1). One innovative approach to **Best Management Practices*** (BMPs) for fertilizer use is termed the *Right Source, Right Rate, Right Time, Right Place* concept, which is also known as 4R nutrient stewardship (International Plant Nutrition Institute 2011) (Figure 4.6).

One "Right Source" method to increase FPUE that has been used successfully is to coat solid fertilizer particles with a copolymer that surrounds the negatively charged phosphate with positive charges, thereby impeding the fixation of phosphate by aluminum, calcium, and magnesium; a product designed for this purpose

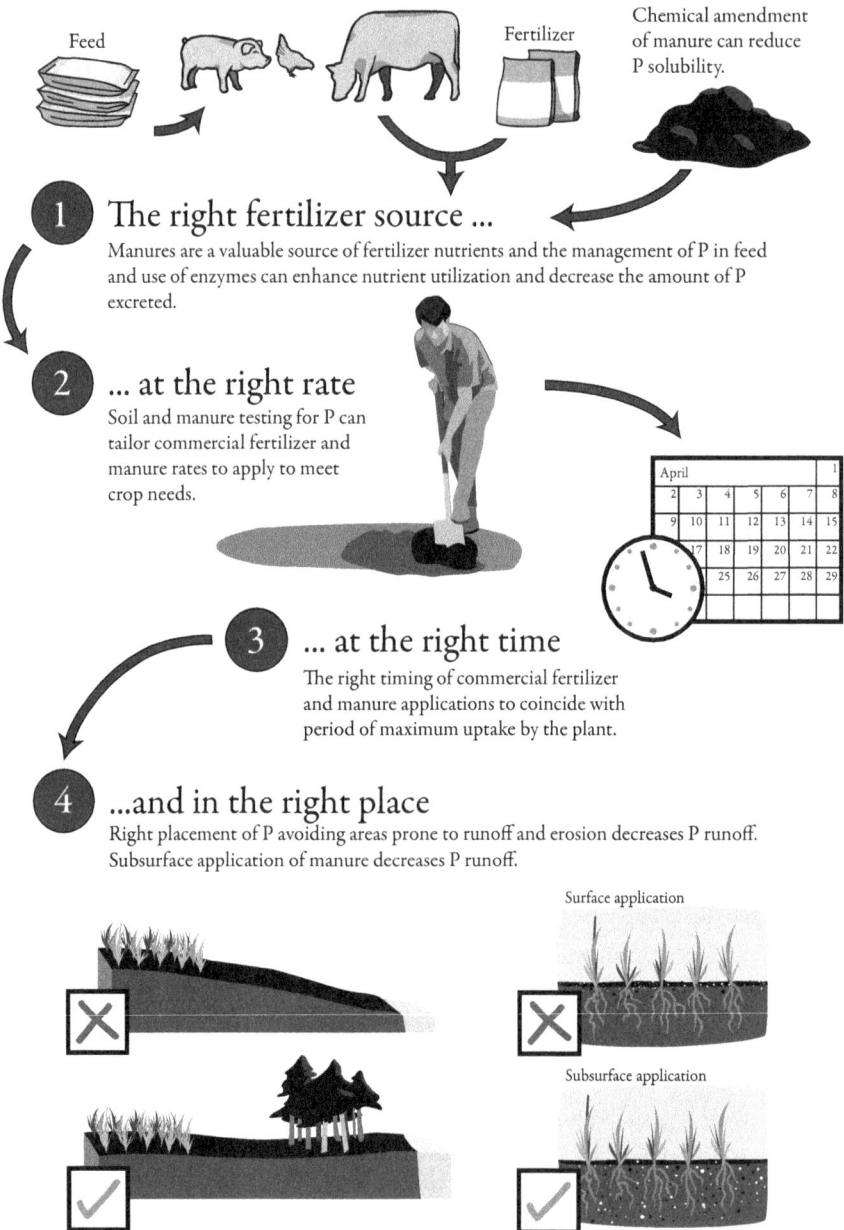

FIGURE 4.6 A possible management solution for fertilizer use is termed the *Right Source, Right Rate, Right Time, Right Place* concept, which is also known as 4R nutrient stewardship (International Plant Nutrition Institute).

is marketed as Avail®. Another means of increasing FPUE is to formulate P fertilizers that include both inorganic and organic P compounds: for example, organic compounds extracted from leonardite (a soft coal) can be blended with phosphoric acid or ammonium phosphate to enhance the bioavailability of fertilizer P; one example of this kind of liquid fertilizer blend is SuperPhos™ (0-50-0). Diammonium phosphate also can be incorporated into organic urea briquettes, and when deeply placed, this P source appears to result in reduced runoff P losses in some agricultural settings (Savant et al. 1997, 1998).

Another source-based method to increase FPUE is the application of liquid P fertilizer directly onto plant leaves, so that P uptake is immediate and unimpeded by cationic soil metals. Such liquid fertilizers are available in a range of P concentrations and formulations with nitrogen and potassium, and give the user options to apply the P such that its absorption by the plant is either through the roots, through the shoot, or both. While this method of application avoids loss of P via precipitation reactions in the soil, it is important for the user to properly dilute these highly concentrated fertilizers in order to avoid "burning" the plant leaves. Alternatively, phosphate rock can be used as a P fertilizer, and the efficient use of this source can be guided by the recently developed **Phosphate Rock Decision Support System*** (PRDSS). The PRDSS is a mathematical model that allows users to compute the relative agronomic efficiency of phosphate rock under a wide range of conditions (Chien et al. 2009). The principal purpose of the PRDSS is to help users to decide whether phosphate rock can be used as a good agronomic substitute for water-soluble fertilizers at the field and farm level (International Atomic Energy Agency 2011).

Traditional soluble P fertilizers include ordinary superphosphate, triple superphosphate, ammonium polyphosphate, monoammonium phosphate, and diammonium phosphate (Young et al. 1985). The use of slow-release P fertilizers such as partially acidulated rock P phosphate on soils other than those with low pH, calcium, and P content have also been evaluated (Hedley et al. 1989; Muchovej et al. 1989). For example, in neutral pH soils it may be possible to apply a heavy initial dressing of finely ground rock phosphate and to include a rotation of fine-rooted legumes in order to generate conditions in the rooting zone that increase the rock phosphate dissolution. Other methods designed to increase soil acidity in the rooting zone include the addition of elemental sulfur (Muchovej et al. 1989), ammonium fertilizers, or organic matter such as animal manure and crop residues (Hedley et al. 1989).

Application rates of P fertilizers typically are first estimated using existing knowledge of the phosphorus demands for the crop being cultivated, but these estimates should then be modified to reflect the amount of P that is already present in the soil, as measured by long-established soil test methods (Cox 1994). For example, Dodd and Mallarino (2005) showed that annual fertilizer P rates of about 15 kg P ha^{-1} maintained near-optimum soil test P levels and corn–soybean yields for Mollisol soils in Iowa. In the case of commercial fertilizer, P applications can easily be tailored to match crop needs and minimize excessive soil P accumulation, because strong economic incentives exist to avoid applying excess fertilizer P. In contrast, applications of P-rich manures have until recently been designed to meet the nitrogen needs of the crop, resulting in the buildup of soil test P above the levels that are needed for optimum crop yields, thereby enhancing the risk of undesirable P losses to runoff (Daverede et al. 2003; Pote et al. 1996; Torbert et al. 2002), and accelerating downstream eutrophication.

However, deciding when to apply fertilizer depends upon logistical and time constraints that are unique to every farm. In principle, the timing of fertilizer application should match as closely as possible the time that it is most needed by the crop being cultivated. Careful timing of fertilizer application minimizes opportunities for surface runoff or leaching below the root zone, and thus should maximize the proportion of applied P that is taken up and used by the crop. As a general rule, fertilizer also should be applied when runoff potential is low; avoiding fertilizer applications within a few days of expected rainfall can help to minimize runoff losses (Sims and Kleinman 2005). Even so, fertilizer rates easily can be tailored to crop needs on the basis of soil test P, minimizing the potential for soil P buildup and P loss in runoff.

The spatial placement of phosphorus is also critical: due to general immobility of P in the soil profile, fertilizer placement is important. Depending on a farm's soils and other environmental factors, **band applications*** of P may or may not be better than **broadcast applications*** of P. For example, Holford (1989) found that fertilizer P effectiveness, as measured by yield response in the first crop, residual effect in the second crop, or cumulative recovery of applied P, was consistently greater for banding at a shallow 5 cm depth than either banding at 15 cm or broadcast applications. Optimal positional availability can also be influenced by crop type, because plant species differ in their responses to soil P availability. Increased root growth and P uptake in the P-fertilized soil layer has been observed for corn (Anghinoni and Barber 1980), soybeans (Borkert and Barber 1985), and wheat (Yao and Barber 1986), relative to plants that were grown in unfertilized soil. In contrast, increased P uptake and yield response by flax plants were obtained when fertilizer P was placed 2–5 cm below the seed, ensuring adequate P levels during early growth (Bailey and Grant 1989).

Much like a checkerboard, most farmers' fields also are not uniform over space with respect to soil properties, topography, and crop growth potential, and these spatial differences can greatly affect crop P fertilization requirements. For example, studies of more than 2000 cores collected from a 58-ha cornfield in Nebraska (Hergert et al. 1994; Peterson et al. 1994) showed that plants grown on about 75 percent of the field would likely respond to P fertilization, while the remaining 25 percent of the farm's total area was P-rich because of soil phosphorus that still remained from long-discontinued livestock operations that existed on this property many decades previously. Farmers with access to modern tools can explicitly incorporate their knowledge of spatial variability into fertilizer management and placement decisions. For example, they can use maps of soil P or past harvests to guide **variable rate fertilizer applications*** (VRA), and can use global positioning system (GPS) technology to optimize their field practices, especially if real-time soil sampling is used to assess the spatial distribution of P needs in any given field. GPS can also be used to turn mobile farm equipment (e.g., planters, sprayers, fertilizer applications, and other tools) on and off as they move across the field, and modern equipment also possess auto steer/parallel tracking capabilities. The use of advanced methods such as these will receive even greater attention in the future as agronomic models are developed that incorporate the field-level information that is increasingly becoming available through precision agriculture technology. Such technology, coupled with soil and nutrient management information, will help to inform the correct rates and methods of P application that are needed to meet crop needs.

Innovations in Plant Genetics to Enhance Plant Phosphorus-Use Efficiency

The fundamental discoveries of Darwin and Mendel established the scientific basis for plant breeding and genetics at the turn of the twentieth century. The recent integration of advances in biotechnology, genomic research, and molecular marker applications with conventional plant breeding practices has created the foundation for molecular plant breeding, an interdisciplinary science that is revolutionizing twenty-first-century crop improvement (Moose and Mumm 2008). Both of these sets of tools are available to help enhance future crop yields by increasing the efficiency with which crop plants take up from external pools and then transform these pools of phosphorus into biomass (phosphorus-use efficiency, expressed here as grams of dry weight produced per gram of P in the harvested plant materials; see Figure 4.2).

The potential for using applied genetics and molecular tools to regulate the within-plant translocation of photosynthate holds great promise for increasing future yields and reducing the P demands of crop production. The partitioning of photosynthetically produced organic carbon between sites of photosynthetic production (the plant's source tissues) and sites of metabolic utilization (the plant's sink tissues) is a major determinant of crop yield (Giaquinta 1983), and therefore should strongly influence PUE. For example, one sine qua non for creating a new generation of more P-efficient crops is the development and use of new methods to enhance plant root systems. Plants respond to nutrient scarcity by allocating a greater proportion of photosynthetically fixed organic carbon to their root systems (Hermans et al. 2006). In response to P-limitation, plants undergo dramatic morphological and architectural changes in their root system in order to increase the absorptive root surface area, and thereby acquire more soil P. These changes include increased extension rates of root growth, an enhanced frequency of lateral root formation, and a higher recruitment of the cells that form root hairs (Abel et al. 2002; Poirier and Bucher 2002). Moreover, P-starved plants such as the model organism *Arabidopsis* (mouse-ear cress) are more sensitive to the plant growth hormone **auxin**,* and this enhanced hormone response has been associated with beneficial modifications in root architecture (Lopez-Bucio et al. 2002; Lopez-Bucio et al. 2003). Many species also respond to sparingly available inorganic phosphate by producing specialized structures called cluster roots, which release substantial amounts of organic acids that acidify the **rhizosphere*** and release bioavailable soil P. In addition, it has been shown that phosphate-starved plants up-regulate a special **H+-PPase enzyme*** (also called vacuolar-type inorganic pyrophosphatase) that couples the energy of pyrophosphate hydrolysis to proton movement across plant membranes (see Box 4.2). Interestingly, genetically modified tomato, rice, and *Arabidopsis* plants with enhanced expression of the H+-PPase have an improved PUE (Yang et al. 2007). These and other efforts to enhance crop yields are very promising and should continue to be pursued aggressively.

For example, tomato plants engineered to over-express AVP1 developed larger shoots, root systems, and fruits than controls when grown under P-deficient conditions (Yang et al. 2007). Root and shoot dry weights of plants grown in the presence of 100 ppm NaH_2PO_4 were, on average, 13 percent and 16 percent higher ($P < 0.01$) in transgenic LeAVP1DOX strain plants than in controls, respectively. Of note, under the same low P conditions, fruit dry weight data and P content per plant were 82 percent and 30 percent higher ($P < 0.01$) than in controls, respectively.

BOX 4.2

The type I H+-PPase gene (AVP1) from *Arabidopsis* has been identified as a gene that enhances plant yield (Li et al. 2005; Gonzalez et al. 2009; Gonzalez et al. 2010) by triggering production of 40–60 percent greater leaf area relative to wild type plants that do not over-express this gene. AVP1 over-expression triggers an enhanced capacity for nutrient uptake in *Arabidopsis*, tomato, and rice plants (Yang et al. 2007). These genetically modified plants out perform wild-type plants when grown under phosphate limitation (Yang et al. 2007; Gaxiola et al. 2011).

Innovations in Animal Genetics to Enhance Phosphorus-Use Efficiency in Livestock

As illustrated in Fig. 4.2, an important facet of phosphorus-use efficiency (PUE) in livestock production is the efficiency with which the animals transform assimilated inorganic P into meat, egg, and milk production (expressed here as grams of dry weight produced per gram of P in the harvested animal materials). Global meat consumption alone is expected to increase by 68 percent over the 2000 base by 2030 (Steinfeld and Gerber 2010 and chapter 7), and this increase in meat consumption is expected to be derived primarily from pig and poultry production. These projected increases in meat consumption in turn can be expected to translate into large increases in the consumption of livestock feeds and thus increase the demand for P.

It is important that we both forecast and mitigate the environmental costs that are associated with livestock production (Pelletier and Tyedmers 2010; Steinfeld and Gerber 2010). One of these potential environmental costs is associated with dietary P consumption: cereal grains and plant protein supplements used in the production of swine and poultry contain from 45 to 79 percent of total phosphorus in the form of *phytate* (Viveros et al. 2000; the chemical structure of this molecule is illustrated in chapter 2). Because swine and poultry lack the enzymes that are necessary to digest phytate, this organic phosphorus compound passes unused and unchanged through the digestive tract, causing P enrichment that results in a lower nitrogen:phosphorus ratio than is ideal for crop growth. If this manure is applied to crops in order to satisfy their nitrogen requirement, phosphorus therefore will be supplied in excess; this excess phosphorus can leach into groundwater or surface waters and result in extensive eutrophication (see chapter 5). To improve P-use efficiency by their livestock, swine and poultry farmers usually include dietary **phytase*** to release inorganic phosphate from a portion of the dietary phytate, thereby decreasing or eliminating the necessity of supplementing the animals' diet with inorganic P (Rao et al. 2009).

There are economic costs associated with these dietary supplements, however, and efforts are being made to reduce or avoid their use via genetic modifications of livestock. For example, to circumvent the need for phytase and phosphate to the diets of swine, Golovan et al. (2001) introduced a transgene that enables the transgenic line of pigs (trademarked as Enviropig™) to grow normally on a cereal grain diet without the inclusion of either supplemental dietary phosphorus or phytase (Figure 4.7; see Box 4.3).

Similar to pigs, poultry lack the capability to digest phytate in cereals despite the presence of weak phytase activity in their intestinal tract (Maenz and Classen 1998). However, it appears that considerable work remains before satisfactory genetic engineering of phytase production can be achieved in poultry. In addition, many fish species can be raised on grain-based diets such as corn, soybean, and wheat. Fish

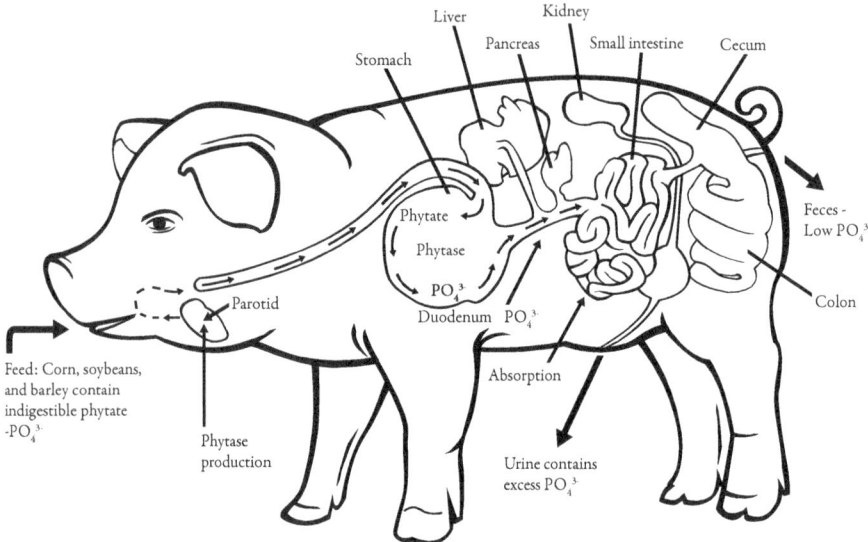

FIGURE 4.7 Schematic showing how the Enviropig™ functions to digest phytate phosphorus in their cereal grain feeds.

BOX 4.3

This phytase expression has global applicability (Forsberg et al. 2003), and comparative studies of analyses of liver and muscle proteins from Enviropigs versus conventional pigs have provided strong evidence that the genetically modified pigs are normal in all respects except the novel phytase capability (Golovan et al. 2008; Hakimov et al. 2009). Since no supplemental phosphorus is added to the Enviropigs' feed ration, much of the cereal grain phosphorus released by phytase is assimilated, resulting in a 35–65 percent reduction in manure phosphorus content. The requirements of the Canadian Environmental Protection Act have already been satisfied for commercial production of Enviropigs in Canada (Government of Canada 2010). However, this transgenic line of pigs will not enter the human food chain until Canadian legislation is met for human food and animal feed consumption, and until approval is obtained from the Food and Drug Administration for commercial production of the Enviropig™ line of pigs in the United States.

---

lack phytase expression, and like pigs, they excrete undigested phytate phosphorus directly into their environment. To reduce the excretion of phosphorus by Japanese Medaka fish, Hostetler et al. (2005) introduced a phytase transgene into their gastrointestinal tissue, and survival of the modified Medaka fish was increased sixfold relative to wild fish when grown on a diet containing phytate as their primary dietary source of phosphorus.

Concerns about phosphorus in animal feeds are not restricted to pigs and fish, however. For example, a substantial increased demand in milk consumption in developing countries is expected to occur by 2030 (Steinfeld and Gerber 2010), resulting in an increase in dairy cattle production and thus a potential increase in phosphorus output as dairy manure. Milk cow forages are typically low in phytate phosphorus (Nelson et al. 1976), and it is essential to supplement the cows' forage diet with a cereal grain in order to improve milk production. Although cattle contain rumen bacteria that produce phytases (Nakashima et al. 2007), Knowlton et al. (2007) have shown that inclusion of phytase in the cows' diet further improves phosphorus utilization. In some situations it may be thus useful to introduce the phytase trait into cows, as has been done for the Enviropig™ line of pigs (Golovan et al. 2001).

BOX 4.4
CHAPTER 4 SUMMARY

- More than tenfold variability exists in the phosphorus content of the world's top 15 agricultural crops, and trends from 1961 to 2009 in P-demands associated with these 15 crops are dominated by sugarcane, wheat, maize, and rice.
- P-demands associated with projected trends in U.S. renewable transportation biofuels are very significant, and indirect and unintentional impacts on food prices from this level of competition with other agricultural P-uses can be expected if major efforts are not made to recycle the P that is contained in the processed plant biomass.
- Improvements in fertilizer phosphorus-use efficiency include the Right Source, Right Rate, Right Time, Right Place concept, which is also known as 4R nutrient stewardship.
- Improvements in the phosphorus-use efficiency of both crop plants and livestock are possible using modern tools of genetic engineering.

REFERENCES

Abel S., C. A. Ticconi, C. A. Delatorre. 2002. Phosphate sensing in higher plants. *Physiologia Plantarum* 115:1–8.

Anghinoni, I., and S. A. Barber. 1980. Predicting the most efficient phosphorus placement for corn. *Soil Science Society of America Journal* 44:1016–1020.

Bailey, L. D., and C. A. Grant. 1989. Fertilizer phosphorus placement studies on calcareous and noncalcareous chernozemic soils: Growth, P-uptake and yield of flax. *Communications in Soil Science and Plant Analysis* 20:635–654.

Baxter, C.A., B. C. Joern, D. Ragland, J. S. Sands, and O. Adeola. 2003. Phytase, high-available phosphorus corn, and storage effects on phosphorus levels in pig excreta. *Journal of Environmental Quality* 32:1481–1489.

Borkert, C. M., and S. A. Barber. 1985. Soybean shoot and root growth and phosphorus concentration as affected by phosphorus placement. *Soil Science Society of America Journal.* 49:152–155.

Carpenter, S. R., N. F. Caraco, D. L. Correll, R. W. Howarth, A. N. Sharpley, and V. H. Smith. 1998. Nonpoint pollution of surface waters with phosphorus and nitrogen. *Ecological Applications* 8:559–568.

Chien, S. H. L., I. Prochnow, and H. Cantarella. 2009. Recent developments of fertilizer production and use to improve nutrient efficiency and minimize environmental impacts. *Advances in Agronomy* 102:267–322.

Childers, D. L., J. Corman, M. Edwards, and J. J. Elser. 2011. Sustainability challenges of phosphorus and food: Solutions from closing the human phosphorus cycle. *BioScience* 61:117–124.

Cox, F. R. 1994. Current phosphorus availability indices: characteristics and shortcomings. In *Soil Testing: Prospects for Improving Nutrient Recommendations,* edited by J. L. Havlin, J. S. Jacobsen, D. F. Lekiam, P. E. Fixen, and G. W. Hergert, 101–114. Soil Science Society of America Special Pub. No. 40, Soil Science Society of America, Madison, WI.

Daverede, I. C., A. N. Kravchenko, R. G. Hoeft, E. D. Nafziger, D. G. Bullock, J. J. Warren, and L. C. Gonzini. 2003. Phosphorus runoff: Effect of tillage and soil phosphorus levels. *Journal of Environmental Quality* 32:1436–1444.

Dodd, J. R., and A. P. Mallarino. 2005. Soil-test phosphorus and crop grain yield responses to long-term phosphorus fertilization for corn-soybean rotations. *Soil Science Society of America Journal* 69:1118–1128.

Ebeling, A. M., L. G. Bundy, M. J. Powell, and T. W. Andraski. 2002. Dairy diet phosphorus effects in phosphorus losses in runoff from land-applied manure. *Soil Science Society of America Journal* 66:284–291.

Forsberg, C. W., J. P. Phillips, S. P. Golovan, M. Z. Fan, R. G. Meidinger, A. Ajakaiye, D. Hilborn, and R. R. Hacker. 2003. The Enviropig physiology, performance, and contribution to nutrient management, advances in a regulated environment: The leading edge of change in the pork industry. *Journal of Animal Science.* 81:E68–E77.

Gaxiola R. A., M. Edwards, and J. J. Elser. 2011. A transgenic approach to enhance phosphorus use efficiency in crops as part of a comprehensive strategy for sustainable agriculture. *Chemosphere* 84(6):840–845.

Giaquinta, R. T. 1983. Phloem loading of sucrose. *Annual Review of Plant Physiology* 34:347–387.

Golovan, S. P., H. A. Hakimov, C. P. Verschoor, S. Walters, M. Gadish, C. Elsik, F. Schenkel, D. K. Y. Chiu, and C. W. Forsberg. 2008. Analysis of Sus scrofa liver proteome and identification of proteins differentially expressed between genders, and conventional and genetically enhanced lines. *Comparative Biochemistry and Physiology—Part D Genomics and Proteomics* 3:234–242.

Golovan, S. P., R. G. Meidinger, A. Ajakaiye, M. Cottrill, M. Z. Wiederkehr, D. Barney, C. Plante, J. Pollard, M. Z. Fan, M. A. Hayes, J. Laursen, J. P. Hjorth, R. R. Hacker, J. P. Phillips, and C. W. Forsberg. 2001. Pigs expressing salivary phytase produce low phosphorus manure. *Nature Biotechnology* 19:741–745.

Gonzalez N., G. T. Beemster, D. Inzé. 2009. David and Goliath: what can the tiny weed Arabidopsis teach us to improve biomass production in crops? *Current Opinion in Plant Biology* 12:157–164.

Gonzalez N., S. De Bodt, R. Sulpice, Y. Jikumaru, E. Chae, S. Dhondt, T. Van Daele, L. De Milde, D. Weigel, Y. Kamiya, M. Stitt, G. T. S. Beemster, and D. Inzé. 2010 Increased leaf size: Different means to an end. *Plant Physiol.* 153:1261–1279.

Government of Canada. 2010. http://www.gazette.gc.ca/rp-pr/p1/2010/2010-02-20/html/notice-avis-eng.html

Hakimov, H. A., S. Walters, T. C. Wright, R. G. Meidinger, C. P. Verschoor, M. Gadish, D. K. Chiu, M. V. Stromvik, C. W. Forsberg, and S. P. Golovan. 2009. Application of iTRAQ to catalogue the skeletal muscle proteome in pigs and assessment of effects of gender and diet dephytinization. *Proteomics* 9:4000–4016.

Hedley, M. J., R. W. Tillman, and G. Wallace. 1989. The use of nitrogen fertilizers for increasing the suitability of reactive phosphate rocks for use in intensive agriculture. In *Nitrogen fertilizer use in New Zealand agriculture and horticulture,* edited by R. E. White and L. D. Currie, 311–320. Occasional Report No. 3, Fertilizer and Lime Research Centre, Massey University, New Zealand.

Hergert, G. W., R. B. Ferguson, C. A. Cotway, and T. A. Peterson. 1994. *Developing accurate nitrogen rate maps for variable rate application.* In Agronomy Abstracts, 399. American Society of Agronomy, Madison, WI.

Hermans, C., J. P. Hammond, P. J. White, and N. Verbruggen. 2006. How do plants respond to nutrient shortage by biomass allocation? *Trends in Plant Science* 11:610–617.

Holford, I. C. R. 1989. Efficacy of different phosphate application methods in relation to phosphate sorptivity in soils. *Australian Journal of Soil Research* 27:123–133.

Hostetler, H. A., P. Collodi, R. H. Devlin, and W. M. Muir. 2005. Improved phytate phosphorus utilization by Japanese Medaka transgenic for the Aspergillus niger phytase gene. *Zebrafish* 2:19–31.

International Atomic Energy Agency. 2011. http://wwwiswam.iaea.org/dapr/srv/en/daprIntroduction

International Plant Nutrition Institute. 2011. http://www.ipni.net/4r

Kimmell, R. J., G. M. Pierzynski, K. A. Janssen, and P. L. Barnes. 2001. Effects of tillage and phosphorus placement on phosphorus runoff losses in a grain sorghum-soybean rotation. *Journal of Environmental Quality* 30:1324–1330.

Knowlton, K. F., M. S. Taylor, S. R. Hill, C. Cobb, and K. F. Wilson. 2007. Manure nutrient excretion by lactating cows fed exogenous phytase and cellulase. *Journal of Dairy Science.* 90:4356–4360.

Lopez-Bucio J., A. Cruz-Ramirez, L. Herrera-Estrella. 2003. The role of nutrient availability in the regulating root architecture. *Current Opinion in Plant Biology* 6:280–287.

Lopez-Bucio J., E. Hernandez-Abreu, L. Sanchez-Calderon, M. F. Nieto-Jacobo, J. Simpson, L. Herrera-Estrella. 2002. Phosphate availability alters architecture and causes changes in hormone sensitivity in the Arabidopsis root system. *Plant Physiology* 129:244–256.

Maenz, D. D., and H. L. Classen. 1998. Phytase activity in the small intestinal brush border membrane of the chicken. *Poultry Science* 77:557–563.

Maguire, R. O., D. A. Crouse, and S. C. Hodges. 2007. Diet modification to reduce phosphorus surpluses: A mass balance approach. *Journal of Environmental Quality* 36:1235–1240.

Maguire, R. O., J. T. Sims, W. W. Saylor, B. L. Turner, R. Angel, and T. J. Applegate. 2004. Influence of phytase addition to poultry diets on phosphorus forms and solubility in litters and amended soils. *Journal of Environmental Quality* 33:2306–2316.

Mankin, K. R., J. K. Koelliker, and P. K. Kalita. 1999. Watershed and lake water quality assessment: An integrated modeling approach. *Journal of the American Water Resources Association* 35:1069–1080.

Mankin, K. R., S. H. Wang, J. K. Koelliker, D. G. Huggins, and J. F. DeNoyelles. 2003. Watershed-lake water quality modeling: Verification and application. *Journal of Soil Water Conservation* 58:188–197.

Moose, S. P., and R. H. Mumm. 2008. Molecular plant breeding as the foundation for 21st Century crop improvement. *Plant Physiology* 147:969–977.

Muchovej, R. M. C., J. J. Muchovej, and V. H. Alvarez. 1989. Temporal relations of phosphorus fractions in an Oxisol amended with rock phosphate and Thiobacillus thiooxidans. *Soil Science Society of America Journal* 53:1096–1100.

Nakashima, B., T. McAllister, R. Sharma, and L. Selinger. 2007. Diversity of phytases in the rumen. *Microbial Ecology* 53:82–88.

National Research Council. 2001. *Nutrient Requirements of Dairy Cattle*, 7th revised edition. National Academy Press, Washington, DC.

Nelson, T. S., L. B. Daniels, J. R. Hall, and L. G. Shields. 1976. Hydrolysis of natural phytate phosphorus in the digestive tract of calves. *Journal of Animal Science* 42:1509–1512.

Pate, R., G. Klise, and B. Wu. 2011. Resource demand implications for US algae biofuels production scale-up. *Applied Energy* 88:3377–3388.

Patterson, P. H., P. A. Moore, Jr., and R. Angel. 2005. Phosphorus and poultry nutrition. In *Agriculture and Phosphorus Management: The Chesapeake Bay,* edited by A. N. Sharpley, 635–682. CRC Press, Boca Raton, FL.

Pelletier, N., and P. Tyedmers. 2010. Forecasting potential global environmental costs of livestock production 2000–2050. *Proceedings of the National Academy of Sciences* 107:18371–18374.

Peterson, T. A., J. S. Schepers, C. Chen, C. A. Cotway, R. B. Ferguson, and G. W. Hergert. 1994. *Interpreting yield and soil parameter maps in the evaluation of variable rate nitrogen applications.* In Agronomy Abstracts, 397. American Society of Agronomy, Madison, WI.

Pierzynski, G. M., and T. J. Logan. 1993. Crop, soil, and management effects on phosphorus soil test levels. *Journal of Production Agriculture* 6:513–520.

Poirier, Y., and M. Bucher. 2002. Phosphate transport and homeostasis in Arabidopsis. In *The Arabidopsis Book,* edited by E. M. Meyerowitz and C. R. Somerville, 1–35. American Society of Plant Biologists, Rockville, MD.

Pote, D. H., T. C. Daniel, A. N. Sharpley, P. A. Moore, Jr., D. R. Edwards, and D. J. Nichols. 1996. Relating extractable soil phosphorus to phosphorus losses in runoff. *Soil Science Society of America Journal* 60:855–859.

Poulsen, H. D. 2000. Phosphorus utilization and excretion in pig production. *Journal of Environmental Quality* 29:24–27.

Rao, D. E. C. S., K. V. Rao, T. P. Reddy, and V. D. Reddy. 2009. Molecular characterization, physicochemical properties, known and potential applications of phytases: An overview. *Critical Review of Biotechnology* 29:182–198.

Savant, N. K., R. G. Menon, and D. K. Friesen. 1997. Transplanting geometry improves timing of uptake of deep point-placed P by rice hills. *International Rice Research News* 22(1):36.

Savant, N. K., and P. J. Stangel. 1998. Urea briquettes containing diammonium phosphate: A potential NP fertilizer for transplanted rice. *Fertility Research* 51:85–94.

Sharpley, A. N., R. W. McDowell, and P. J. A. Kleinman. 2001. Phosphorus loss from land and water: Integrating agricultural and environmental management. *Plant Soil* 237:287–307.

Sharpley, A. N., and S. Rekolainen. 1997. Phosphorus in agriculture and its environmental implications. In *Phosphorus Loss from Soil to Water*, edited by H. Tunney, O. T. Carton, P. C. Brookes, and A. E. Johnston, 1–54. CAB International Press, Cambridge, England.

Sharpley, A. N., P. J. A. Withers, C. Abdalla, and A. Dodd. 2005. Strategies for the sustainable management of phosphorus. In *Phosphorus; Agriculture and the Environment*, edited by J. T. Sims and A. N. Sharpley, 1069–1101. *Am. Soc. Agron. Monograph 46*. American Society of Agronomy, Madison, WI.

Simpson, T. W., A. N. Sharpley, R. W. Howarth, H. W. Paerl, and K. R. Mankin. 2008. The new gold rush: Fueling ethanol production while protecting water quality. *Journal of Environmental Quality* 37:318–324.

Sims, J. T., and P. J. A. Kleinman. 2005. Managing agricultural phosphorus for environmental protection. In *Phosphorus, Agriculture and the Environment*, edited by J. T. Sims and A. N. Sharpley, 1021–1068. *Am. Soc. Agron. Monograph 46*. American Society of Agronomy, Madison, WI.

Sims, J. T., and A. N. Sharpley. 2005. *Phosphorus; Agriculture and the Environment. Am. Soc. Agron. Monograph*. Agronomy 46. American Society of Agronomy, Madison, WI.

Smith, S. J., A. N. Sharpley, W. A. Berg, J. W. Naney, and G. A. Coleman. 1992. Water quality characteristics associated with Southern Plains grasslands. *Journal of Environmental Quality* 21:595–601.

Steinfeld, H., and P. Gerber. 2010. Livestock production and the global environment: Consume less or produce better? *Proceedings of the National Academy of Sciences* 107:18237–18238.

Torbert, H. A., T. C. Daniel, J. L. Lemunyon, and R. M. Jones. 2002. Relationship of soil test phosphorus and sampling depth to runoff phosphorus in calcareous and noncalcareous soils. *Journal of Environmental Quality* 31:1380–1387.

USDA—National Agricultural Statistics. 2010. *U.S. Department of Agriculture—National Agricultural Statistics Service*. 2010. National statistics for corn. Washington, DC.

Valk, H., J. A. Metcalf, and P. J. A. Withers. 2000. Prospects for minimizing phosphorus-excretion in ruminants by dietary manipulation. *Journal of Environmental Quality* 29:28–36.

Viveros, A., C. Centeno, A. Brenes, R. Canales, and A. Lozano. 2000. Phytase and acid phosphatase activities in plant feedstuffs. *Journal of Agricultural Food Chemistry* 48:4009–4013.

Wu, Z., L. D. Satter, A. J. Blohowiak, R. H. Stauffacher, and J. H. Wilson. 2001. Milk production, phosphorus excretion, and bone characteristics of dairy cows fed different amounts of phosphorus for two or three years. *Journal of Dairy Science* 84:1738–1748.

Yang H, J. Knapp, P. Koirala, D. Rajagopal, W. A. Peer, L. Silbart, A. Murphy, and R. Gaxiola. 2007. Enhanced phosphorus nutrition in monocots and dicots over-expressing a phosphorus-responsive type I H+-pyrophosphatase. *Plant Biotechnology Journal* 5:735–745.

Yao, J., and S.A. Barber. 1986. Effect of one phosphorus rate placed in different soil volumes on P uptake and growth of wheat. *Communications in Soil Science and Plant Analysis* 17:819–827.

Young, R. D., D. G. Westfall, and G. W. Colliver. 1985. Production, marketing, and use of phosphorus fertilizers. In *Fertilizer Technology and Use*, 3rd ed., 323–376. Soil Science Society of America, Madison, WI.

5

Phosphorus in Urban and Agricultural Landscapes

P IS FOR PRESERVATION

Shelby H. Riskin, Gaston Small, Robert Mikkelsen, Genevieve Metson, Anna Bateman, James Cooper, Ola Stedje Hanserud, Philip M. Haygarth, Cecilia Laspoumaderes, Michelle McCrackin, Sonya Remington

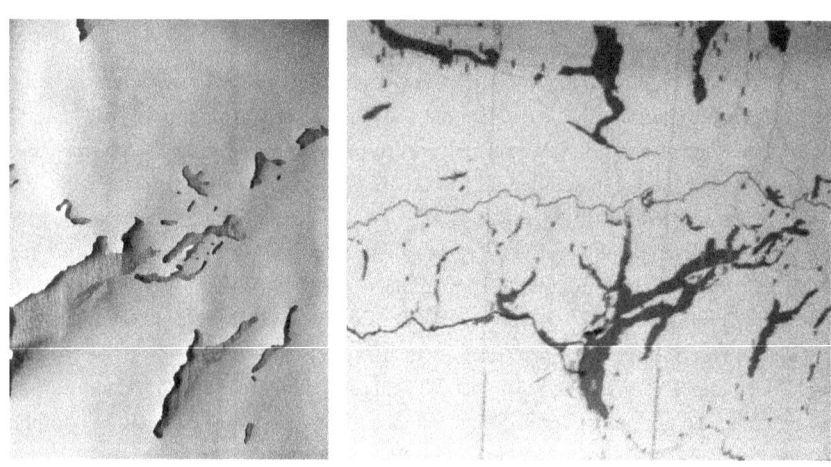

Erik L. Peterson; *Pristine Lack: Patagonian Lakes;* Hand-cut paper, 31" x 82" 2011

Erik L. Peterson
Pristine Lack: Patagonian Lakes

Scientific Collaborator:
Cecilia Laspoumaderes, PhD student,
Universidad del Comahue-CONICET, Argentina

Description of Artwork:
Dr. Laspoumaderes's work in Argentina's Patagonia region reveals the awesome power of species survival in inhospitable, nutrient-poor lakes, whose untouched beauty and clarity is caused by that deficiency. It is striking to me that the very thing that makes the Patagonian lakes pristine, clear, and beautiful (the absence of phosphorus) actually denotes a lack. This paradox made for a very interesting opportunity for my "Pristine Lack: Patagonian Lakes" drawing. Defining lack (the deficiency of nutrients) in terms of clarity (the transparency of the lakes) is manifest in the field of drawing as "negative space." Negative space is an artistic and graphic design term that denotes the space between marks that the artist makes on paper, or can be understood as the white space between black letters on a page. Using an X-Acto knife, I have incised this lack, or "negative space," directly into the paper to create an aerial view of the region of Patagonia in which Dr. Laspoumaderes works. The white expanse of the paper stands in for land and the lakes are represented by their absence from the page. When lit, the "clear" lakes will show on the wall behind the paper as pure light and will also offer glimpses of viewers on either side of the work, illustrating our own complicity in the care and destruction of aquatic ecosystems.

About the Artist:
Erik L. Peterson's work intervenes in the interstitial spaces of public discourse, activating both the physical fabric of urban topography and investigating contemporary mythology. Artworks like "Self-Serve Soft-Serve," a municipal pipe that pumps frozen yogurt, introduce play into the public space of the city, while projects like "Qeej Hero," a video game that reanimates a 4,000-year-old Hmong musical instrument, show his commitment to collaboration. Peterson has presented his artwork and design projects internationally at a wide range of academic and industry conferences, spanning the fields of graphic design (AIGA, 2008), ethnography (EPIC, 2009), and performance studies (PSi, 2010).

BOX 5.1
CHAPTER 5 OBJECTIVES

- We present case studies that argue that although temporal and spatial aspects of P sustainability vary widely, appropriate local efforts should be made to maximize P-use efficiency and recycling of waste both within the system and between linked ecosystems.
- We also demonstrate that P budgets (inputs and outputs) provide a useful indicator of long-term sustainability, but a zero P budget is not always desirable or achievable in a short time frame.
- We recommend management changes for ecosystems that currently have a large surplus of P to facilitate better use of this limited natural resource.

Phosphorus (P) is essential for plant growth, and for most of human history agricultural production has been supported by local recycling of organic phosphorus (see chapters 1 and 2 for more details). The dramatic increases in global agricultural production in the twentieth century, however, were made possible in part by increasing inputs of inorganic P fertilizer (mined phosphate rock) (Tilman et al. 2001, FAO 2009).

As part of this increasing P-use in food production, humans have tripled the annual amount of P moving through the global environment compared to natural flows (Smil 2000). This increase in environmental P includes increased erosion and increases in urban wastewater flows and manure, but the most key factor is the increased use of mined rock phosphorus as fertilizer. More than a third of this added fertilizer P is accumulating in agricultural soils, both in occluded and biologically available forms (Bennett 2001). Another fraction is entering aquatic ecosystems, either at the beginning of the human food chain from agricultural runoff and leaching, or at the end via wastewater effluent, contributing to **eutrophication*** (Schindler 2006, Carpenter and Bennett 2011).

Recent attention has been drawn to how much alteration of the P cycle the globe can withstand without the crossing of irreversible thresholds of environmental degradation (Rockstrom et al. 2009, Carpenter and Bennett 2011, Townsend and Porder 2011). Based on oceanic eutrophication, the threshold from which oceans would not recover was defined as preindustrial P flows multiplied by ten. Using freshwater eutrophication to define the boundary, however, it has been argued that we have already crossed an irreversible threshold from P misuse (Carpenter and Bennett 2011).

In preindustrial times, the flux of P from land to surface waters resulted primarily from natural weathering of P-containing minerals. Now, runoff from fertilized

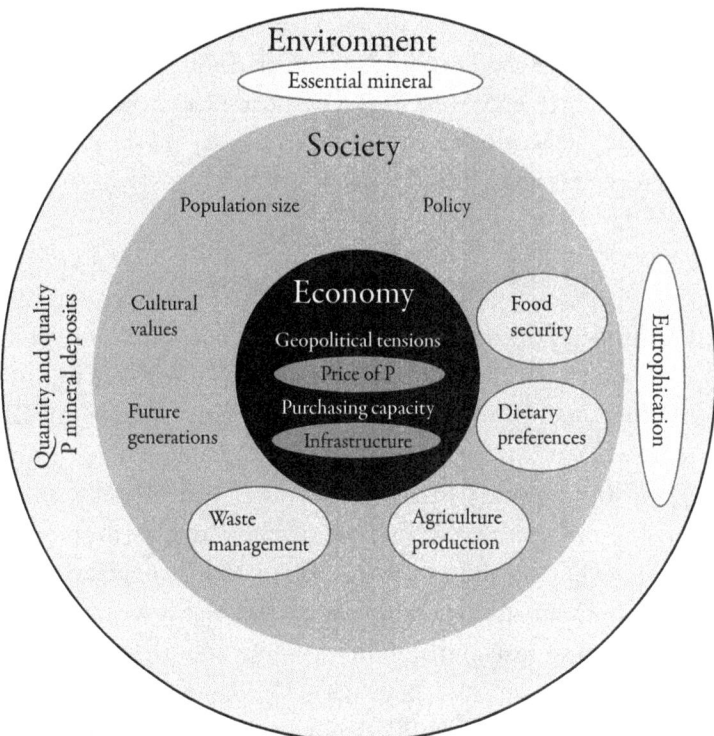

FIGURE 5.1 Phosphorus management in relation to sustainability thinking. The image illustrates the 3 large considerations of sustainability: environment (white sphere), society (gray sphere), and economy (black sphere) as a nested model where environment encompasses society, and economy is part of society. Words in each sphere are elements that contribute to the phosphorus management problem. Circled words in each sphere are important to P management in urban and agricultural landscapes.

fields and concentrated animal feeding operations, erosion, and discharge from sewage treatment plants, have increased P loading to surface waters. Globally, river networks now transport approximately 9 Tg P to coastal areas each year (Seitzinger et al. 2010). Over 80 percent of dissolved inorganic P delivered to these coastal areas is attributed to human activities (Seitzinger et al. 2010). Scenarios of future P loading to coastal areas predict that these rates will further increase, driven primarily by sewage and agriculture (Seitzinger et al. 2010).

There is significant regional variation in fluxes of P, reflecting differences in human activities and climate: About 66 percent of the total P flux from land to coastal areas comes from tropical latitudes, while 28 percent is from temperate latitudes (estimated from Seitzinger et al. 2010). These differences are driven both by the natural environment (ecosystem nutrient limitation and soils) as well as by human factors such as land use (the extent of agriculture and urban areas) and degree of development (the existence of water treatment infrastructure) and lead to variable **P balances**[*] and impacts. Improving the efficiency of P usage and recycling along the human food chain is important and urgent for sustainability (Figure 5.1). In many cases, however, improving P-use also involves economic trade-offs that must be overcome. Successful approaches to improving P-use must consider environmental, societal, and economic heterogeneity to increase the sustainability of urban and agricultural systems across regions (Figure 5.1).

ENVIRONMENTAL VARIABILITY

Many factors influence the outcomes of P management. While all soils hold reservoirs of P, the total quantity and **bioavailability*** is controlled by a number of environmental factors (Jenny 1941), including climate (Porder and Chadwick 2009), parent material (Soderberg and Compton 2007), soil age (Walker and Syers 1976), and dust inputs (Chadwick et al. 1999). Additionally, soils have the ability to retain P in forms that are biologically unavailable (for more information see, chapter 2). This ability to retain P, or the **P sorption capacity*** of soils, varies with a number of soil properties. Old, highly weathered soils with high concentrations of iron and aluminum oxides and high clay content typically have high sorption capacity (Sanchez 1976, Zhang et al. 2005), as do **alkaline soils*** where P precipitates with calcium carbonate ($CaCO_3$) (Iyamuremye and Dick 1996). Thus, added fertilizer P will be immobilized in biologically unavailable forms in some soils, while remaining accessible to plants in others. Soils with high sorption capacities require higher P inputs to maintain crop growth and will be able to receive higher P inputs before leaching to surface waters is possible.

Soils used for agriculture fall along a spectrum of both P availability and P saturation. Many traditional agricultural regions like those of North America and Europe are dominated by soils, like Mollisols and Alfisols, with relatively large pools of labile P and relatively low sorption capacity (FAO 2011). In contrast, many tropical regions, often countries with actively developing agricultural economies, are dominated by tropical and highly weathered soils, like Oxisols and Ultisols, with high P sorption capacities and relatively small reservoirs of available P (FAO 2011). These soils require continuous and large P inputs to attain and maintain yields comparable to those in more temperate regions, adding an additional challenge and cost to developing successful agricultural systems in these regions.

The sorption capacity of any soil can be saturated, with different amounts of added P necessary to saturate the binding capacity of different soils. The P sorption capacity can also be altered. For example, increasing the pH or the organic matter content of soils can decrease the P sorption capacity of soils as well as increasing the

P available to plants (Guppy 2005, Zhang et al. 2005). Balancing P inputs with crop demands and soil reservoirs is possible and achieved through periodic soil testing and precision fertilizer applications (chapter 4).

LAND USE AND MANAGEMENT VARIABILITY

Agriculture

Since the widespread availability of soluble P fertilizer, inputs of P to agricultural land have exceeded P removed through plant uptake, grazing, and harvest on a global scale (Bouwman et al. 2009). However, both current soil P stocks and the rate of change of soil P balances (accumulation or deficit) vary regionally, often depending on economic status. In many developed countries where agricultural soils have received regular inputs of P fertilizer since the second half of the twentieth century, fertilizer P application by farmers has now largely stabilized, occurring in tandem with improved agronomic practices and more direct and careful P management (Vitousek et al. 2009, MacDonald et al. 2011).

In some rapidly developing countries, agricultural intensification is now expanding the use of commercial fertilizers. Intensification of existing cropland through increased fertilization can reduce land conversion, and P additions are necessary to build soil fertility and increase yields. However, overuse is causing some regions, like China, to accumulate large P surpluses and to experience increasing eutrophication of surface waters (Vitousek et al. 2009, MacDonald et al. 2011). In much of Africa, where the cost of importing P has remained prohibitive, agricultural soils are heavily depleted of P, as evidenced by low crop yields across the region (Sanchez 2002).

Livestock-based agricultural systems affect P balances mainly through animal manures. In low-density and low-input livestock systems, most P is recycled: animals ingest P in grazed plant material and return 98 percent of this consumed P to the grazed area as manure and urine (Drangert 1998). The remaining small fraction of P is exported from the landscape in the animal biomass.

Increasingly common across much of the world are concentrated livestock operations, where high densities of livestock are raised in confinement and are fed imported feed rather than grazing or otherwise feeding from the local environment. In such operations nutrients are imported from a wide area in the form of feed, and the resulting animal waste is concentrated in a relatively small area near the farm. Costs of transporting animal waste prohibit moving it far, so most is spread on adjacent cropland, sometimes at high concentrations that exceed crop nutrient requirements. These cropping systems can accumulate P surpluses, facilitating the creation of nutrient excess "hotspots" that may contribute to nutrient runoff to aquatic systems (Whalen and Chang 2001). These systems do not currently facilitate efficient P recycling and reuse.

Urban Systems

In urban systems in developed countries, P that exits the human food chain as excreta, other organic waste, and in detergents is transported to wastewater treatment plants (see chapter 6 for detailed discussion). There, a fraction of P (depending on the specific procedures used) is recovered as sludge and buried in landfills; the remainder is discharged into surface waters. To achieve substantial removal of P, additional technology must be incorporated into wastewater systems, which may be cost prohibitive. Many developing countries do not have wastewater treatment infrastructure and P (in untreated sewage) may be discharged directly to waterways. On a global average, sewage processing removes only about 28 percent of P from wastewater, though this will likely increase (Van Drecht et al. 2009). The remaining P is generally discharged to surface waters. If sewage treatment technology does not keep pace with rates of urbanization and increased connectivity to sewage systems, P inputs to surface waters could double in the next several decades (Van Drecht et al. 2009).

Receiving Ecosystems

Phosphorus is exported from agricultural or urban landscapes through streams and rivers, eventually reaching lakes, reservoirs, and coasts. Some of these P inputs are diffuse, like runoff from saturated agricultural soils or erosion, while others, such as that from concentrated livestock operations and urban wastewater outflows, are point sources that are more easily identified and controlled.

Excessive inputs of P to surface waters can have a stimulatory effect on aquatic vegetation, driving eutrophication (Carpenter et al. 1998, Fisher et al. 1999). It is possible to both limit the eutrophication of unimpacted systems and to mitigate eutrophication in those already impacted by reducing P and other nutrient inputs from the receiving ecosystems, but this can be costly and recovery slow (Carpenter et al. 1998, Schindler 2006). Limiting export from both point and non-point sources can more effectively limit the impact of nutrient enrichment.

In most cases, agricultural and urban systems are not managed for P directly. Instead, farmers attempt to maximize yields with fertilizers, livestock managers minimize costs by dealing with animal waste as economically as possible, and urban systems treat waste as their infrastructures and technologies allow. The following case studies illustrate some of the challenges associated with P in urban and agricultural landscapes as these systems are currently managed. We identify where these systems misuse and/or lose P to surrounding environments, and we offer opportunities for improvement. We use illustrations of P balances from agricultural and urban systems with different environments, land use, and infrastructure.

Improving P-use will not be possible with a single or simple solution. Regional differences in P-use and its consequences require consideration of site-specific solutions, and economic incentives may be required to encourage direct management of P. However, many widely applicable opportunities for improvement in P management exist along the human food chain. For example, soil P, particularly organic soil P, is a reservoir of P that could potentially be more effectively made available to crops. Recycling P from animal and human waste will reduce losses to water that can contribute to eutrophication and preserve natural P reserves. We can also improve and are improving P management by basing P inputs on soil tests and soil properties, reducing erosion, avoiding over-application of livestock waste to croplands, and installing and maintaining wastewater treatment with appropriate capacities. Here, we look at P-use globally across urban and agricultural systems to characterize the current state of P management along the human food chain.

CASE STUDIES

Crop-Based Agriculture

Adequate P supplies were lacking in most of the long-cultivated lands in eighteenth- and nineteenth-century North America and Western Europe, with no readily available sources accessible to overcome crop deficiencies. Animal manure additions did not provide the correct proportions of essential nutrients for the rapidly expanding agricultural industry (Mikkelsen and Bruulsema 2005).

Beginning in the mid–nineteenth century, ground animal bones were treated with sulfuric acid to form single superphosphate [$Ca(H_2PO_4)_2 \cdot H_2O$], the first commercial P fertilizer. Leading brands of bone-derived P fertilizer averaged approximately 2 percent P, far lower than those commonly used today (up to 24 percent P) (Mikkelsen and Bruulsema 2005).

The solubility of phosphate rock is generally quite low. The direct application of phosphate rock to fields is usually only recommended for very acidic soils with rock that is finely ground and well-distributed in the soil (Chien et al. 2010). It is most effective for longer-term and perennial crops (Chien et al. 2010). Treated phosphate rock is the main source of P fertilizer today (Mikkelsen and Bruulsema 2005).

There is a common misperception that farm nutrients are self-sustaining. Despite the very best management and internal cycling of nutrients on farms, all crop and animal agriculture bring unavoidable P losses, mostly through the export of harvested products. Any farming system that does not replace the lost or harvested nutrients will eventually become depleted. Periodic soil testing, affordable fertilizer supplies, and careful fertilizer application will reduce P imbalances, both surpluses and deficits.

Sub-Saharan Africa

The Green Revolution drove impressive yield increases across much of the world, but large expanses of sub-Saharan Africa continue to struggle to produce adequate food supplies. Many of the region's soils are highly weathered with high P sorption capacity and low nutrient status prior to any cultivation (Sanchez 1980, Sanchez 2003), making nutrient requirements higher than many agricultural regions. Nutrient inputs, however, including P, lag behind those of any other global region, with an average of 9 kg nutrients ha^{-1} compared with 100 kg ha^{-1} in East Asia and 230 kg ha^{-1} in Western Europe (IMPHOS 2011). Years of harvesting crops without nutrient replacement have depleted already low native soil fertility. There is an urgent need to improve the nutrient supply in sub-Saharan Africa (Sanchez 2005, Annan 2010, IMPHOS 2011). However, fertilizer prices, lack of capital, poorly developed markets, and inadequate infrastructure currently prevent this from happening (Sanchez 2002, Sanchez 2005).

Large yield increases across sub-Saharan Africa have been shown possible following increased P inputs. Through increased nutrient inputs and management, maize yields can more than double at both the village and country scale (Sanchez 2009, IMPHOS 2011). For example, in Uasin Gishu, Kenya, experimental fields applying 26 kg P ha^{-1} increased maize yields from 0.2 t ha^{-1} to 7.3 t ha^{-1} (IMPHOS 2011). In this situation, the annual P budget went from a deficit of approximately 2 kg P ha^{-1} yr^{-1} to a surplus of almost 19 kg P ha^{-1} yr^{-1}. To increase yields across sub-Saharan Africa it will be necessary to continue to accumulate P surpluses for many years before soil P concentrations are increased such that the constraints of nutrient limitation are consistently overcome.

United States and Canada Phosphorus Balances

The most current results from the International Plant Nutrition Institute (IPNI) summarizing soil nutrient concentrations in North American cropland analyzed approximately 4.4 million soil samples collected in the fall of 2009 or the spring of 2010 (IPNI 2010a). Phosphorus concentrations varied widely among as well as within the individual states of the United States. In general, the northern Great Plains had the lowest soil P concentrations, and the far eastern region had the highest. Using these soil P concentrations, the percentage of samples where yield loss would be anticipated without P additions were calculated (Figure 5.2). For example, in North Dakota, 86 percent of the samples predict yield losses without P additions. In contrast, in New Jersey only 13 percent of analyses predict any loss. Such an analysis highlights not only that not all soils in developed countries are over-fertilized with P, but also that periodic soil testing is important, to establish the need for additional nutrients on a regional and field-specific basis.

A recent mass balance of primary agronomic nutrients (N, P, and K) to quantify nutrient removal by crops, fertilizer applied, and excreted and recoverable manure nutrients in the United States showed large spatial differences in P balances, but also identified a number of general trends (Figure 5.3, IPNI 2010b). The national P balance in 2007 remained positive at 230,000 t P. Over 20 years (1987–2007), however, the total amount of P removed in harvested crops increased faster than fertilizer P inputs

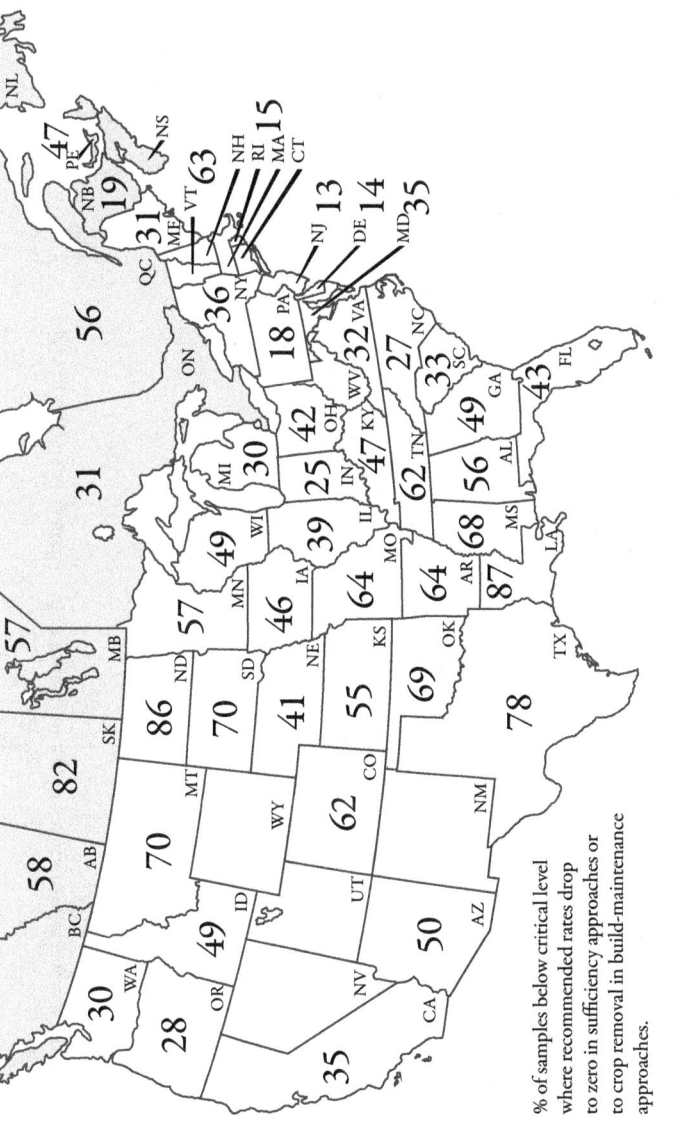

FIGURE 5.2 Percent of soil samples with less P than the critical concentration for major crops in 2010 (IPNI 2010a).

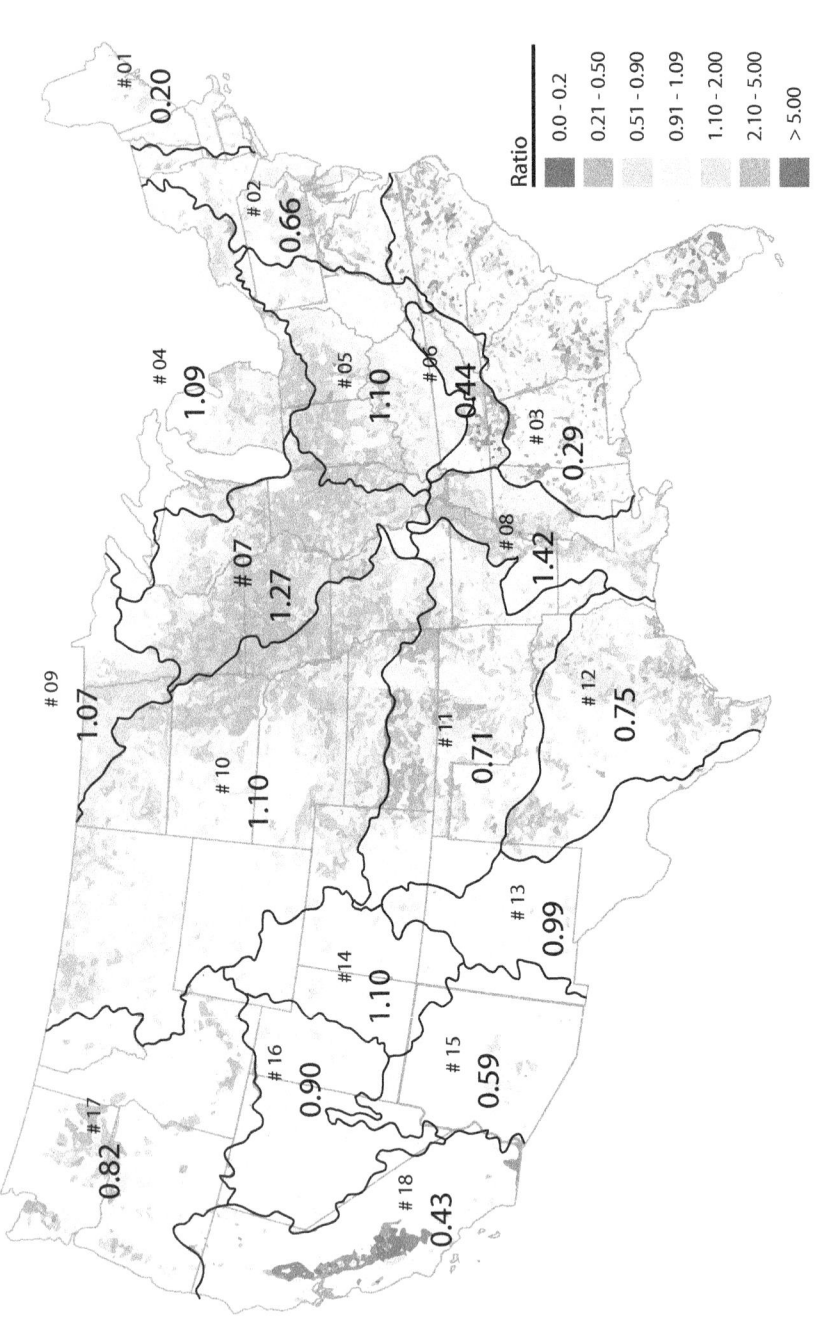

FIGURE 5.3 Estimated P removal to use ratio in agriculture by hydrologic region in 2007 (IPNI 2010b).

and recoverable manure P. During this period, the average annual increase in crop removal was 33,000 t P, while P fertilizer use increased at only 13,000 t P yr^{-1}.

The P balance for individual states differ greatly. For example, Illinois went from a positive P balance in 1987 to a deficit in 2007, such that crop removal exceeded P inputs by over 50 percent. Several eastern states and California tend to have a more positive P balance than the rest of the country (Figure 5.3).

In much of North America and Western Europe, following years of fertilizer application, soil P concentrations no longer limit crop production. This strategy of gradually building up the P concentration in soil to a satisfactory level (building phase) reduces the risk of nutrient deficiencies and maximizes economic return. Most crop advisors recommend that soils that are very high in P should not be fertilized until the P concentrations decline with crop removal. This requires periodic soil testing and record keeping. For example, McCollum (1991) reported that without P fertilizer additions, 16 to 18 years of corn and soybean production would be needed to deplete soil P concentrations back to the agronomic threshold in a North Carolina cropping system. Similarly, Eghball et al. (2003) reported that soil P concentrations in Nebraska would decrease to levels prior to intense cultivation within 10 years of cropping if no additional P was added.

Intensive Soy in Mato Grosso, Brazil

Large-scale crop-based agricultural operations are relatively new to the Amazon region. The dominant soils in the region are highly weathered, acidic, and nutrient poor (Quesada et al. 2011). They were long thought to be unable to sustain crop-based agriculture (Sanchez and Logan 1992, Quesada et al. 2011).

One large farm in Mato Grosso, Brazil, has been growing soybeans since 2003. Fields were converted from pasture to soybeans between 2003 and 2008 and the farm now has approximately 400 km^2 in soybean production each year. The farm adds approximately 50 kg inorganic P fertilizer ha^{-1} yr^{-1}, of which approximately 18 kg P ha^{-1} yr^{-1} are removed in the harvested crop (Riskin, et al. 2013). Of what remains, a very small amount may be lost to wind and water erosion or to leaching; the soil infiltration rate is quite high, however, suggesting that rates of overland flow and thus water erosion are very low (Hayhoe et al. 2011). These highly weathered soils have high P sorption capacity, binding P in forms that are not bioavailable on short timescales (Sanchez 1976). All soybean seed is exported, though crop residues are returned to the soil.

Because of the soil's capacity to sorb P in forms that are unavailable in the short term, repeated additions of P are necessary to maintain high yields. In the lowland tropics that are dominated by these soils, the short-term consequences of P surplus will likely not be increased nutrient runoff and eutrophication as has been the case in many agricultural regions. Instead, the challenge will be building adequate bioavailable P for crop growth, while considering the constraints imposed by the farm budget and the global P supply.

Livestock-Based Agriculture

Intensive livestock production in many countries typically involves large-scale feeding operations, where feed is imported, the target products (e.g., poultry, milk, etc.) are exported, and manure accumulates on-site or nearby. Ideally, P sustainability in livestock production entails maximizing efficiency in converting P from feed into salable products, and then maximizing local recycling of the manure.

Three fundamental challenges impede the sustainable use of P in livestock production. First, P-use efficiency at the farm-scale requires enhancing the P-use efficiency of individual animals, which is presently quite low. Between 40 and 90 percent of P in corn and soybean feed is in the form of **phytate*** [also called phytic acid] (Harland and Oberleas 1996). This P-rich compound is largely indigestible to **monogastric*** animals that do not produce the phytase enzyme required for phytate hydrolysis and utilization. This necessitates the addition of inorganic P supplements to the feed to meet the animal growth requirement. As a result, the P concentration in manure is usually elevated.

Technological solutions to this low P efficiency problem include (1) adding phytase to animal feed (Nys et al. 1996), (2) developing feed crops containing lower levels of phytate and higher levels of digestible P (Ertl et al. 1998), and (3) developing transgenic animals capable of producing phytase (i.e., the "Enviropig"; see chapter 4 and Forsberg et al. 2003). Phytase supplements are now commonly used in animal feed, whereas regulatory and social barriers to the development of transgenic livestock have precluded the commercial use of this technology.

Second, confined livestock production results in a concentration of manure-based P into a relatively small area. For example, a typical broiler production facility (a concentrated animal production facility raising chicken for meat) that imports 2500 Mg of feed annually (Tarkalson and Mikkelsen 2003) required production from ~250 ha of land to balance nutrient additions (based on 2009 average corn yields, NRCS, 2011). Because the nutrient content of manure is low relative to inorganic fertilizer (especially for liquid manure), the cost of transport rapidly exceeds its economic value as fertilizer. Therefore, livestock manure tends to be repeatedly applied to fields in the vicinity surrounding the farm, where P accumulates (Whalen and Chang 2001). Soil characteristics and management practices determine the amount of P that can be retained in the soil before P loss in runoff or to groundwater.

Technological innovations are decreasing the cost of manure transport (Hadrich et al. 2010), allowing manure to be transported over a greater area. However, the fundamental problem remains that animal feed is transported from thousands of miles to feedlots, but nutrients from livestock and manure cannot be economically returned to these farms. Thus, the spatial extent of our food distribution network precludes the recycling of P from manures on a large scale.

Third, the ratio of N to P in animal manure is not in balance with the requirements for plant growth. For example, cattle feedlot manure typically has an N:P ratio between 4 and 5, compared to crop requirements of 6 to 8 (ILOC 1995). Manure has been traditionally applied to meet crop requirements for N, leading to over-application of P. Conversely, the application of manure based on crop requirements for P may minimize P accumulation, but will require additional inputs of inorganic N.

TABLE 5.1
PHOSPHORUS (P) BUDGET FROM A BROILER FARM AND DAIRY FARM IN NORTH CAROLINA, USA (TARKALSON AND MIKKELSEN 2003).

Broiler farm	(98 ha)	Dairy farm	(113 ha)
P inputs	kg yr^{-1}	**P inputs**	kg yr^{-1}
Feed and concentrates	15,000	Concentrates	3700
Bedding	14	Bedding	2
Chicks	75	Fertilizer P	60
Total inputs	15,089	**Total inputs**	3762
P outputs	kg yr^{-1}	**P outputs**	kg yr^{-1}
Broilers	6940	Milk	2500
Mortalities	109	Mortality	40
Corn and soybeans	1660	Calves	7
		Crops	66
		Straw	7.5
Total outputs	8709	**Total outputs**	2621
Surplus in soil	6380	Surplus in soil	1141

Broiler Farm, North Carolina, United States

A P budget calculated for a broiler farm in North Carolina (Tarkalson and Mikkelsen 2003) saw annual P inputs total 15,089 kg, with feed accounting for 99 percent (Table 5.1). Of this, 46 percent was converted into chickens sold off the farm. The chicken litter was spread over an adjacent 98 ha, so an additional 11 percent was converted to secondary products (corn and soybeans) that were exported from the farm. The remaining 42 percent of P inputs were accounted for as surplus in soil, corresponding to a loading rate of 65 kg P ha^{-1} yr^{-1}. Because the sandy coastal plain soil has low capacity for P sorption, the potential for P loss through runoff and leaching in this system is relatively high.

Dairy Farm, North Carolina, USA

On a 113-ha dairy farm also located in North Carolina (Tarkalson and Mikkelsen 2003), P inputs totaled 3762 kg annually (Table 5.1). Of this, 68 percent was converted into milk and animals exported from the farm, and 2 percent of total P inputs were recycled (as calves, crops, and straw that remained on site). The remaining 30 percent went to soil surplus, corresponding to a loading rate of 20 kg P ha^{-1} yr^{-1}. The authors identified additional opportunities for P recycling through manure application to soybean and alfalfa production, which would increase the inputs recycled to 13 percent and decrease the annual soil surplus to 20 percent of total inputs.

TABLE 5.2
PHOSPHORUS (P) BUDGET FROM A DAIRY FARM AND SHEEP FARM IN THE UK (HAYGARTH ET AL. 1998).

Dairy farm	(57 ha)	Sheep farm	(841 ha)
P inputs	kg yr^{-1}	**P inputs**	kg yr^{-1}
Atmosphere	13	Atmosphere	93
Fertilizer	910	Fertilizer	403
Straw Bedding	12	Feed and concentrates	159
Concentrates	1540		
Total inputs	**2475**	**Total inputs**	**655**
P outputs	kg yr^{-1}	**P outputs**	kg yr^{-1}
Calves	31	Lambs	61
Milk	890	Wool	0.53
Losses to water	57	Ewes and hogs	56
		Burning heather	84
		Losses to water	216
Total outputs	**978**	**Total outputs**	**418**
Surplus in soil	1497	Surplus in soil	237

Intensive Dairy Farm, UK

In a phosphorus budget for an intensive dairy farm in the UK, Haygarth et al. (1998; Table 5.2) found the total annual P input to the farm to be 2480 kg P, primarily as cattle feed (62 percent) and fertilizer (37 percent). Each year 890 kg P were exported as milk and 31 kg P as calves, making P utilization efficiency 37 percent. Recycling occurred as manure was deposited directly on pasture during grazing, and was collected and spread to land during the housed period. Of the 3390 kg P consumed annually as grass, silage, and mineral supplements, 70 percent was recycled from the herd within the farm. Of the input, 64 percent was either stored within the system or lost. Small P losses occurring by leaching and runoff (estimated to be between 1 and 3 kg P ha^{-1} yr^{-1}; see chapter 2 for information on P concentrations and eutrophication in surface waters), leaving a surplus of 24 to 26 kg P ha^{-1} yr^{-1} in the soils.

Innovations to improve P efficiency could target reductions of inputs to the system. Manipulation of cattle diets can reduce P intake by 20 to 50 percent without adverse effects to milk production (Dou et al. 2003). Precision feeding has been shown to reduce fecal P concentrations by 33 percent (Cerosaletti et al. 2004). Reduced fertilizer and manure application may also reduce losses by 0.21 kg P ha^{-1} yr^{-1} in intensive dairy systems at zero cost (Haygarth et al. 2009). However, P in manure generated on-site exceeds crop requirements and excess will need to be exported off the farm to prevent additional soil accumulation.

Hill Sheep Farm in the UK

For a hill sheep farm, which produces sheep for meat and wool through grazing on marginal lands, in the UK, a phosphorus budget by Haygarth et al. (1998; Table 5.2) found the total annual P input to the farm to be 655 kg, primarily as fertilizer for improved grassland (62 percent) and imported feeds (24 percent). Only 118 kg was exported in animal products, making a P utilization efficiency of 18 percent. Recycling occurred via consumption of vegetation and depositing of manure and plant litter on the soil. Of the total 909 kg P consumed annually, 83 percent was taken from native pasture and 87 percent was deposited as manure; 70 percent of the P in above-ground vegetation was returned to the soil as plant litter. Wasteful output occurred due to burning (13 percent of input) and losses to water (33 percent of input), leaving a small surplus in soils of 0.28 kg P ha^{-1} yr^{-1}. Because soil accumulation was low, innovations to improve P efficiency should target reductions of losses to water; more than half the quantity imported as fertilizer was lost in this way. Shallow spiking or **subsoiling*** to disrupt compacted soil layers could reduce the soil component of P loss by 50–70 percent (Cuttle 2007), thus improving plant uptake and reducing the need for imported fertilizer.

Urban Environment

Although cities occupy a small percentage of the earth's surface, they are hotspots of P accumulation due to human activity and excreta. Cities concentrate P from food (crop and animal) that is transported from around the world, resulting in waste products rich in P. The concentration of food in cities is accompanied by P losses associated with transport and processing between farm and fork (not examined here; see, e.g., Smit et al. 2009). In addition to external food sources from conventional agriculture, urban agriculture is an important source of food in some cities (Redwood 2009), providing 15 to 20 percent of global food output and contributing to the urban P cycle (Smit et al. 1996).

Treatment of high-P waste (e.g., food waste and human and animal excreta) differs between developed and developing countries. In developed countries, urban food waste not disposed through the sewage system ends up in landfills, is incinerated, or is collected for composting. In developing countries, centralized solid waste management is not as common. Excreta (urine and feces) in developed countries is typically collected and transported through sewage systems to a wastewater treatment facility, removing much of the P (ignoring septic tanks). In many developing countries, untreated waste is discharged to waterways or used directly for irrigation of crops. Economic development and cultural preferences of an urban population also affect P cycling and differences exist within cities (e.g., for planned upper- and middle-class areas vs. unplanned peri-urban areas and slums).

In addition to wasting much food before it reaches our mouths (Childers et al. 2011), humans excrete 98 percent of ingested P (Drangert 1998). Diet also affects the amount of P excreted. Cordell et al. (2009) estimated that a vegetarian excretes some 0.3 kg P yr^{-1} while a person consuming a meat-based diet excretes 0.6 kg P yr.$^{-1}$ Global sewage emissions contained 1.3 Tg of P in 2000 and this will continue to increase as the global population grows (Van Drecht et al. 2009).

In order to manage P more sustainably in cities, we should reduce preconsumption waste, encourage P-efficient diets, and ensure P is properly treated after consumption. Urban waste should be recycled back to agricultural production to reduce water quality impacts and decrease dependence on new P resources. The degree of potential P recycling from urban wastewater will depend on the specific treatment method (Kvarnström and Nilsson 1999; Kvarnström et al. 2000).

Phoenix, Arizona, United States

The Greater Phoenix Area in the Sonoran Desert of Arizona is one of the larger metropolitan areas in the United States, with a population of approximately 4 million. This urban, arid ecosystem is a strong sink for P, where annual inputs are much greater than outputs, accumulating approximately 6000 t P yr^{-1} or 9.4 kg ha^{-1} yr^{-1}, both in soils and landfills (Metson et al. 2012). Known pools sum to 3800 t (59 kg ha^{-1}). Inputs are dominated by food (3800 Mt P) and fertilizer (1900 t P), while exports include agricultural products, wastewater, and paper (totaling 700 t P).

Phoenix has several large internal P flows related to food production, landscaping, and waste management. Flows related to food production dominate the movement of P, while the built environment and population density correspond to the most concentrated pools of P in the city. The biophysical characteristics of this arid environment (notably **calcareous soils***, low rainfall, and few freshwater bodies) have limited losses of P from the system, translating into P accumulation and low eutrophication risk. Human decisions about waste management (e.g., landfills), water management (e.g., recycling of effluent), food purchasing, as well as landscaping and urbanization all influence the P balance. The proximity of other land uses, particularly agricultural production, also shapes the opportunities for P recycling.

The modification of local hydrology in response to concerns about water scarcity in such an arid system has played a major role in internal P cycling, where treated wastewater is routinely applied to agricultural fields and urban landscaping rather than being exported from the city. Water resource scarcity motivates wastewater reuse, not P management. The city may be suited for small-scale, household, or community recycling of human excreta and large volumes of organic waste produced as yard trimmings and food scraps. Smaller-scale recycling may also be compatible with the city's landscape, where agriculture is a mosaic in the urban development, instead of surrounding the city.

Norway

National-level P fluxes in Norway that are linked to wastewater treatment facilities are derived from statistical data for 2009. In 2009, 83 percent of 4.8 million Norwegians were served by a wastewater treatment plant (WWTP) with a capacity larger than 50 **population equivalents*** (pe) (Statistics Norway 2011a). Seventy-nine percent of the population lived in densely populated areas (defined as an area with at least 200 inhabitants and a maximum distance of 50 meters between houses; Statistics Norway 2011b). It is assumed that all persons living in densely populated areas are served by a WWTP of >50 pe and that the national average treatment efficiency for this size facility also holds true for the served urban population.

There were 3300 t P in wholesale food and beverages for urban areas in 2009, while 2500 t P reached treatment facilities through sewer systems. The loss of P along the way consisted of food waste that never entered the wastewater collection system and leakages of wastewater from the sewer system (5 percent of transported P). Industry and service add to the P in wastewater, as did the trucking and emptying of wastewater into the system from non-sewered houses with septic tanks. In 2009 the average P retention in WWTPs of more than 50 pe was 70.6 percent (Statistics Norway 2011a), discharging 750 t of P into waterways and retaining 1800 t P in biosolids for the urban population. Of the retained **biosolid,*** 56 percent was used in agriculture, 22 percent for urban greening, and 12 percent deposited in landfills, including as soil cover (Statistics Norway 2011a). However, P from sewage sludge substituted little mineral fertilizer P (Bøen and Grønlund 2008), and farmers' main motivations for receiving sewage sludge did not often include the P content, but rather concentrations of **lime*** (when sludge is limed), N, and organic matter.

Receiving Ecosystems

Beyond well-developed urban areas with long-standing infrastructure like Phoenix and Norway, challenges exist to managing wastewaters with increasing industrialization, population density, and urbanization. Population growth and rising affluence increase water consumption, resulting in increasing volumes of wastewater entering local and regional water bodies (streams, rivers, lakes, coastal areas). Much of this wastewater remains untreated, with more than 2.6 billion people living in areas without adequate treatment facilities (Millennium Ecosystem Assessment 2005), leading to nutrient and chemical loading, which can drive freshwater eutrophication (Carpenter and Bennett 2011, Schindler 2006).

Sewage treatment is expensive, as it has extensive infrastructure requirements, and it is difficult to determine the appropriate intensity of effluent treatment as well as the appropriate treatment capacity in areas with increasing urban populations (Álvarez-Vázquez et al. 2011). The nutrient composition of wastewater effluent can drive the degradation of receiving water bodies (Akpor and Muchie 2010), so appropriate treatment of wastewater is often the final opportunity for preventing nutrient loading of surface waters by urban waste.

Lake Nahuel Huapi, Patagonia, Argentina

Lake Nahuel Huapi is located in the Nahuel Huapi National Park (southern Argentina), where no significant industrial or agricultural activities exist and the only potential source for anthropogenic nutrient loading is the urban population (Diaz et al. 2007). The lake has a glacial origin, a maximum depth of 440 m, an area of 530 km^2, and the residence time of water in the lake is 12 years. The landscape is mountainous, surrounded by well-developed forests, and extends to the margins of the Patagonian Steppe.

The city of San Carlos de Bariloche, on the southeastern edge of the lake, is one of the largest cities in the region and has been expanding rapidly with increasing migration from cities like Buenos Aires (25 percent annually between 1991 and 2010). Prior to the construction of a WWTP in 1996, city sewage was discharged directly into the lake without treatment. With the construction of the plant, wastewater from 70 percent of the city population is now treated with an **activated sludge process*** to reduce P and N concentrations (Alemanni 2006). The plant generates about 5000 m^3 biosolid yr^{-1} that is turned to compost and sold regionally.

The arrival of tourists drives seasonal variation in effluent loading, overflowing plant capacity during some parts of the year and leading to occasional discharges of untreated sewage to the lake (Madariaga 2007). Despite these overflows, Lake Nahuel Huapi remains relatively unaffected by wastewater effluent. The WWTP reduces anthropogenic inputs, and nutrient concentrations across the lake have not yet shown differences in nutrient loading through time or space (Alemanni 2006). Recent organic matter and nutrient inputs to the lake have been mainly in a dissolved form, promoting uptake by algae and fast dispersal in the water column.

Lake Nahuel Huapi remains classified as an oligotrophic lake despite increasing population pressure (Pedrozo 1997, Alemanni 2006). To maintain this status, however, there remains a need to increase the capacity of the WWTP, both to ensure adequate sewage treatment during the high tourist season and to keep pace with continuing population growth. In the longer term, developing cities like San Carlos de Bariloche must develop infrastructure to capture and potentially recycle P waste to prevent degradation of receiving surface waters.

Wapato Lake, Washington, United States

Wapato Lake is a 14 ha urban lake located in Tacoma, Washington, and, in contrast to Lake Nahuel Huapi, has been heavily impacted by anthropogenic nutrient loading. It is a natural lake formed approximately 15,000 years ago during the retreat of the Frazier Ice Sheet. Today, its watershed includes a major interstate highway, shopping center, and residential areas. The lake is **polymictic*** and shallow (2 m deep).

Nutrient inputs have driven eutrophication in Wapato Lake. Phosphorus inputs include internal recycling from sediments and external loading from storm water and waterfowl. In contrast to a stratified lake, Wapato Lake is a prime candidate for eutrophication, as occasional mixing by wind resuspends sediment P, making it available to algae, and calm periods during which algal growth causes oxygen concentrations to become depleted further the process (Scheffer and Jeppesen 2007, Tetra Tech 2008).

Human impacts on Wapato Lake began in the late nineteenth century with the development of a public park. During this time the lake became a popular swimming area and was stocked with trout and waterfowl. In 1942 swimmers began developing rashes and it was discovered that a stormwater sewer had been transporting sewage directly to the lake. Data collected in the 1970s indicated that eutrophication was well under way. Beginning with a dredging operation in 1936, water-quality remediation and management efforts have been unsuccessful in returning the lake to its pre-eutrophied condition (Tetra Tech 2008). These efforts included dredging lake sediments, flushing with freshwater, copper sulfate treatments to reduce plankton populations, alum treatments to remove P, and building a constructed wetland at the lake's inlet (Tetra Tech 2008, Welch and Jacoby 2001).

Despite these management efforts, Wapato Lake remains eutrophic. This case study demonstrates the difficulty in time and expense associated with restoring water quality of aquatic ecosystems receiving excessive P inputs.

Sustainable Solutions and Future Directions

The examples presented here show how environmental variability, land use, stakeholder objectives, and infrastructure influence the vulnerability of systems to inefficient P-use. With increasing fertilizer prices and increasing awareness of P pollution risks, future management strategies should focus more directly on P: on minimizing P inputs, accessing alternative sources of P, recycling animal and human wastes, and ultimately limiting inputs to aquatic ecosystems. This must be done while sustaining food production and operating within the constraints of global markets and government budgets. When adopting sustainable solutions, regional variability must be incorporated such that changes to local policy and practice are successful at addressing each area's unique P challenges.

Crop-Based Agriculture

Crop-based agriculture is the largest consumer of fertilizer P (Smil 2000) and offers many opportunities to improve P management. In addition to soil testing and careful management of P inputs, existing reservoirs of soil P could be exploited better. For example, incorporating organic matter into soils, particularly soils with high P sorption capacity, can both reduce sorption of added P and increase P uptake by plants (Guppy et al. 2005). Eliminating other soil constraints, such as excess acidity or an inadequate supply of essential nutrients, also improves the recovery of P. Precision placement of P in the soil makes relatively immobile nutrients more accessible to plant roots. Not all cropping situations require high-solubility P sources, allowing lower grades of P rock to be used for fertilizer. With improved recycling and reclamation of animal and human waste, fertilizer P inputs derived from apatite may also be increasingly replaced with recycled products (see chapter 3 for detailed discussion about the mineral aspects of P).

Livestock-Based Agriculture

One of the biggest challenges to improving P management in livestock-based agriculture is reducing and reclaiming P in animal waste and dispersing this resource over large areas to avoid saturating soils in a localized area (Whalen and Chang 2001). Manure-based P should be used to the benefit of crop-based agricultural systems at appropriate rates of application. Reducing the P concentration in waste is possible with improved animal digestion of P by changes in feed and by genetic modifications in animals (Forsberg et al. 2003).

The present system of intensive livestock production makes sustainable nutrient management challenging. Manure is heavy and costly to transport, so methods of reclamation and dispersal will require the implementation of technologies that facilitate concentration and affordable transport (Hadrich et al. 2010) and the creation of nutrient markets so that the use of manure fertilizers becomes economically viable.

Urban Environment

The recycling of P from human waste is possible and technologies to make this viable in urban environments are increasingly available. It is estimated that by 2050 there will be 1.5 Mt P available from reclamation of urine generated in urban areas (Mihelcic 2011). If recovered, this would satisfy approximately 20 percent of the world's annual P demands (Mihelcic 2011). This presents an opportunity for innovation, particularly in developing nations where population growth and expansion of urban populations is most rapid—in many cases in areas currently without sewage and wastewater treatment facilities. Sewage treatment systems are long-lived and costly to replace, so creating and installing new treatment systems is challenging and urgent. Sweden has committed to reducing the P in wastewater by 60 percent by 2015, and other developed nations are likewise working on such systems (Syers et al. 2011).

Receiving Ecosystems

Eutrophication in aquatic ecosystems receiving excess nutrient inputs is a symptom of P surplus and inadequate management. It is possible to mitigate water quality with restoration efforts like **riparian buffers,*** dredging, and other techniques, but these efforts are costly and eutrophication can be persistent (Carpenter et al. 1998, Schindler et al. 2006). Source reduction should be an important part of any P management plan (Carpenter and Bennett 2011). A strategy to minimize P inputs to surface water should start with reduction of P export from agricultural and urban landscapes.

BOX 5.2
CHAPTER 5 SUMMARY

- Phosphorus plays an essential role in agroecosystems and urban areas and it must be carefully balanced between extremes of deficiency and surplus. Each situation has unique opportunities and constraints for achieving better utilization of P.
- Building adequate soil concentrations of bioavailable P is essential for sustainable food production, but the target concentrations are site-specific and may change over time. Monitoring soil P concentrations and avoiding loss are essential components of a successful management plan.
- The P balance for many urban areas and for large animal production systems is excessive. The current systems are not designed to use P efficiently and new models should be considered when possible. Efforts should be made to improve P recovery and recycling where feasible.
- Water bodies receiving inputs of P behave differently. Efforts should be focused first on P source reduction before remediation attempts are made.

REFERENCES

Akpor, O. B., and M. Muchie. 2010. Bioremediation of polluted wastewater influent: Phosphorus and nitrogen removal. *Scientific Research and Essays* 5:3222–3230.

Alemanni, M. E. 2006. *Estado trófico del lago Nahuel Huapi en relación con el crecimiento poblacional urbano en San Carlos de Bariloche.* Universidad del Comahue, San Carlos de Bariloche.

Alvarez-Vázquez, L. J., N. García-Chan, A. Martínez, and M. E. Vázquez-Méndez. 2011. SOS: A numerical simulation toolbox for decision support related to wastewater discharges and their environmental impact. *Environmental Modelling and Software* 26:543–545.

Annan, K. 2010. *Global Conference on Agriculture, Food Security, and Climate Change.* The Hague, Netherlands.

Bennett E. M., S. R. Carpenter, N. F. Caraco. 2001. Human impact on erodable phosphorus and eutrophication: A global perspective. *Bioscience* 51:227–234.

Bøen, A. and Grønlund, A. 2008. Phosphorus resources in waste—closing the loop? In *Phosphorus Management in Nordic-Baltic Agriculture—Reconciling Productivity and Environmental Protection*, edited by G. H. Rubæk, 102–106. NJF Report, Vol. 4 Nr. 4. Uppsala.

Bouwman, A. F., A. H. W. Beusen, and G. Billen. 2009. Human alteration of the global nitrogen and phosphorus soil balances for the period 1970–2050. *Global Biogeochemical Cycles* 23:GB0A04.

Carpenter, S. R., N. F. Caraco, D. L. Correll, R. W. Howarth, A. N. Sharpley, and V. H. Smith. 1998. Nonpoint pollution of surface waters with phosphorus and nitrogen. *Ecological Applications* 8:559–568.

Carpenter, S. R., and E. M. Bennett. 2011. Reconsideration of the planetary boundary for phosphorus. *Environmental Research Letters* 6:014009.

Cerosaletti, P. E., D.G. Fox, L.E. Chase. 2004. Phosphorus reduction through precision feeding of dairy cattle. *Journal of Dairy Science* 87:2314–2323.

Chadwick, O. A., L. A. Derry, P. M. Vitousek, B. J. Huebert, and L. O. Hedin. 1999. Changing sources of nutrients during four million years of ecosystem development. *Nature* 397:491–497.

Chien, S. H., L. I. Prochnow, and R. L. Mikkelsen. 2010. Agronomic use of phosphate rock for direct application. *Better Crops* 94:21–23.

Childers, D. L., J. Corman, M. Edwards, and J. J. Elser. 2011. Sustainability challenges of phosphorus and food: Solutions from closing the human phosphorus cycle. *Bioscience* 61:117–124.

Cordell, D., J.O. Drangert, and S. White. 2009. The story of phosphorus: Global food security and food for thought. *Global Environmental Change* 19:292–305.

Cuttle, S. P., C. J. A. Macleod, D.R. Chadwick, D. Scholefield, P. M. Haygarth, P. Newell Price, D. Harris, M. A. Shepard, B. J. Chambers, and R. Humphrey. 2007. *An Inventory of Methods to Control Diffuse Water Pollution from Agriculture.* Department for Environment, Food, and Rural Affairs (DEFRA) project ES0203, London, United Kingdom.

Díaz, M., F. Pedrozo, C. Reynolds, and P. Temporetti. 2007. Chemical composition and the nitrogen-regulated trophic state of Patagonian lakes: Limnology of Temperate South America. *Limnologica—Ecology and Management of Inland Waters* 37:17–27.

Dou, Z., J. D. Ferguson, J. Fiorini, J. D. Toth, S. M. Alexander, L. E. Chase, C. M. Ryan, K. F. Knowlton, R. A. Kohn, A. B. Peterson, J. T. Sims, and Z. Wu. 2003. Phosphorus feeding levels and critical control points on dairy farms. *Journal of Dairy Science* 86:3787–3795.

Drangert, J. 1998. Fighting the urine blindness to provide more sanitation options. *Water SA* 24:1–8.

Eghball, B., J. F. Shananan, G. E. Varvel, and J. E. Gilley. 2003. Reduction of high soil test phosphorus by corn and soybean varieties. *Agronomy Journal* 95:1233–1239.

Ertl D. S., K. A. Young, and V. Raboy. 1998. Plant genetic approaches to phosphorus management in agricultural production. *Journal of Environmental Quality* 27: 299–304.

FAO. 2009. *How to Feed the World in 2050*. Food and Agriculture Organization of the United Nations (FAO), Rome, Italy.

FAO. 2011. *Soil Suborders Mapping Data*. Food and Agriculture Organization of the United Nations (FAO). http://www.fao.org/geonetwork/srv/en/metadata.show?id=14116, accessed February 2013.

Fisher, T. R., A. B. Gustafson, K. Sellner, R. Lacouture, L. W. Haas, R. L. Wetzel, R. Magnien, D. Everitt, B. Michaels, and R. Karrh. 1999. Spatial and temporal variation of resource limitation in Chesapeake Bay. *Marine Biology* 133:763–778.

Forsberg, C.W., J. P. Phillips, S. P. Golovan, M. Z. Fan, R. G. Meidinger, A. Ajakaiye, D. Hilborn, and R. R. Hacker. 2003. The Enviropig physiology, performance, and contribution to nutrient management, advances in a regulated environment: The leading edge of change in the pork industry. *Journal of Animal Science* 81:E68–E77.

Guppy, C. N., N. W. Menzies, P. W. Moody, and F. P. C. Blamey. 2005. Competitive sorption reactions between phosphorus and organic matter in soil: a review. *Australian Journal of Soil Research* 43:189–202.

Hadrich J. C., T. M. Harrigan, and C. A. Wolf. 2010. Economic comparison of liquid manure transport and land application. *Applied Engineering in Agriculture* 26:743–758.

Harland, B. F., and D. Oberleas. 1996. Phytic acid complex in feed ingredients. In *Phytase in Animal Nutrition and Waste Management,* edited by M. B. Coelho and E. T. Kornegay, 69–75. BASF Corp., Mount Olive, NJ.

Haygarth, P. M., H. ApSimon, M. Betson, D. Harris, R. Hodgkinson, and P. J. Withers. 2009. Mitigating diffuse phosphorus transfer from agriculture according to cost and efficiency. *Journal of Environmental Quality* 38:2012–2022.

Haygarth, P. M., P. J. Chapman, S. C. Jarvis, and R. V. Smith. 1998. Phosphorus budgets for two contrasting grassland farming systems in the UK. *Soil Use and Management* 14:160–167.

Hayhoe, S. J., C. Neill, S. Porder, R. McHorney, P. Lefebvre, M. T. Coe, H. Elsenbeer, and A. V. Krusche. 2011. Conversion to soy on the Amazonian agricultural frontier increases streamflow without affecting stormflow dynamics. *Global Change Biology* 17:1821–1833.

IMPHOS. 2011. *IMPHOS Phosphate Newsletter*. World Phosphate Institute (IMPHOS), Casablanca, Morocco.

Intensive Livestock Operations Committee. 1995. *Code of practice for the safe and economic handling of animal manures. Agdex 400/ 27–2. Alberta Agriculture*, Food and Rural Development, Edmonton, Alberta, Canada.

International Plant Nutrition Institute (IPNI). 2010a. *Soil Test Levels in North America. 30-3110*. International Plant Nutrition Institute, Norcross, GA.

International Plant Nutrition Institute (IPNI). 2010b. *Nutrient Use Geographic System for the U.S. 30-3270*. International Plant Nutrition Institute, Norcross, GA.

Iyamuremye, F. and R. P. Dick. 1996. Organic Amendments and Phosphorus Sorption by Soils. *Advances in Agronomy* 56:139–185.

Jenny H. 1941. *Factors of Soil Formation: A System of Quantitative Pedology*. McGraw-Hill, New York.

Kvarnström, E., C. Morel, J. C. Fardeau, J. L. Morel, and S. Esa. 2000. Changes in the phosphorus availability of a chemically precipitated urban sewage sludge due to different dewatering processes. *Waste Management and Research* 18:249–258.

Kvarnström, E., and M. Nilsson. 1999. Reusing phosphorus: Engineering possibilities and economic realities. *Journal of Economic Issues* 33:393–402.

MacDonald, G. K., E. M. Bennett, P. A. Potter, and N. Ramankutty. 2011. Agronomic phosphorus imbalances across the world's croplands. *Proceedings of the National Academy of Sciences* 108:3086–3091.

Madariaga, M. 2007. *Interacción entre ambiente y población en San Carlos de Bariloche*. INTA, Bariloche.

McCollum, R.E. 1991. Buildup and decline in soil phosphorus: 30 year trends on a Typic Umprabuult. *Agronomy Journal* 83:77–85.

Metson, G., R. D. Hale, E. Iwaniec, J. Cook, J. R. Corman, C. Galletti, and D. Childers. 2012. Phosphorus in Phoenix: a budget and spatial representation of phosphorus in an urban ecosystem. *Ecological Applications* 22(2): 705–721.

Mihelcic, J. R., L. M. Fry, and R. Shaw. 2011. Global potential of phosphorus recovery from human urine and feces. *Chemosphere* 84:832–839.

Mikkelsen, R.L., and T.W. Bruulsema. 2005. Fertilizer use for horticultural crops in the U.S. during the twentieth century. *Hort Technology* 15:24–30.

Millennium Ecosystem Assessment (MA). 2005. *Ecosystems and Human Well-Being: Conditions and Trends*. Island Press, Washington, DC.

NRCS. 2011. Plant nutrient content database. *Natural Resources Conservation Service (NRCS)*. http://plants.usda.gov/npk/main, accessed October 2012.

Nys Y., D. Frapin, and A. Pointillart. 1996. Occurrence of phytase in plants, animals and microorganisms. In *Phytase in Animal Nutrition and Waste Management*, edited by M. B. Coelho and E.T. Kornegay, 213–236. BASF Corp., Mount Olive, NJ.

Pedrozo, F., R. Alcalde, and M. Manuel. 1997. *Estado trófico del Lago Nahuel Huapi y estimación preliminar de su posible evolución*. Centro Regional Universitario Bariloche, Universidad Nacional del Comahue, San Carlos de Bariloche.

Porder S., O. A. Chadwick. 2009. Climate and soil-age constraints on nutrient uplift and retention by plants. *Ecology* 90: 623–636.

Quesada, C. A., J. Lloyd, L. O. Anderson, N. M. Fyllas, M. Schwarz, and C. I. Czimczik. 2011. Soils of Amazonia with particular reference to RAINFOR sites. *Biogeosciences* 8:1415–1440.

Redwood, M., editor. 2009. *Agriculture in Urban Planning*. Earthscan / IDRC, London.

Riskin, S. H., S. Porder, C. Neill, A. M. Figueira, C. Tubbesing, and N. Mahowald. 2013. The fate of fertilizer phosphorus along a chronosequence of soybean fields on the Amazonian agricultural frontier. *Philosophical Transactions of Royal Society B* 20120154. http://dx.doi.org/10.1098/rstb.2012.0154

Rockstrom, J., W. Steffen, K. Noone, A. Persson, F. S. Chapin, E. F. Lambin, T. M. Lenton, M. Scheffer, C. Folke, H. J. Schellnhuber, B. Nykvist, C. A. de Wit, T. Hughes, S. van der Leeuw, H. Rodhe, S. Sorlin, P. K. Snyder, R. Costanza, U. Svedin, M. Falkenmark, L. Karlberg, R. W. Corell, V. J. Fabry, J. Hansen, B. Walker, D. Liverman, K. Richardson, P. Crutzen, and J. A. Foley. 2009. Planetary boundaries: exploring the safe operating space for humanity. *Ecology and Society* 14:32–65.

Sanchez, P., G. Denning, and G. Nziguheba. 2009. The African Green Revolution moves forward. *Food Security* 1:37–44.

Sanchez, P. A. 1976. *Properties and Management of Tropical Soils*. John Wiley, New York.

Sanchez, P. A. 2002. Ecology—Soil fertility and hunger in Africa. *Science* 295:2019–2020.

Sanchez, P. A., and T. J. Logan. 1992. Myths and Science about the Chemistry and Fertility of Soils in the Tropics. In *Myths and Science of Soils of the Tropics*. Soil Science Society of America, Inc., Madison, WI.

Sanchez, P. A., C. A. Palm, and S. W. Buol. 2003. Fertility capability soil classification: a tool to help assess soil quality in the tropics. *Geoderma* 114:157–185.

Sanchez, P. A., and M. S. Swaminathan. 2005. Hunger in Africa: the link between unhealthy people and unhealthy soils. *Lancet* 365:442–444.

Sanchez, P. A., and G. Uehara. 1980. Management consideration for acid soils with high phosphorus fixation capacity. *The Role of Phosphorus in Agriculture,* 471–514. American Society of Agronomy, Madison, WI.

Scheffer, M., and E. Jeppesen. 2007. Regime shifts in shallow lakes. *Ecosystems* 10:1–3.

Schindler, D. W. 2006. Recent advances in the understanding and management of eutrophication. *Limnology and Oceanography* 51:356–363.

Seitzinger, S., E. Mayorga, A. F. Bouwman, C. Kroeze, A. H. W. Beusen, G. Billen, G. Van Drecht, E. Dumont, B. M. Fekete, J. Garnier, and J. A. Harrison. 2010. Global river nutrient export: A scenario analysis of past and future trends. *Global Biogeochemical Cycles* 24 GB0A024.

Smil, V. 2000. Phosphorus in the environment: Natural flows and human interferences. *Annual Review of Energy and the Environment* 25:53–88.

Smit, A. L., P. S. Bindraban, J. J. Schröder, J. G. Conijn, and H. G. van der Meer. 2009. *Phosphorus in agriculture: global resources, trends and developments*. Plant Research International. B.V., Wageningen.

Smit, J., A. Ratta, and J. Nasr. 1996. *Urban agriculture: Food, jobs and sustainable cities*. United Nations Development Programme (UNDP), New York.

Soderberg, K. and J. S. Compton. 2007. Dust as a nutrient source for fynbos ecosystems, *South Africa*. *Ecosystems* 10:550–561.

Statistics Norway. 2011a. *Folkemengde, etter kjønn og tettbygd/spredtbygd strøk (K)*. Statistikkbanken, Emne: 02 Befolkning. Tabell: 05212. http://statbank.ssb.no/statistikkbanken/ . Accessed February, 2011.

Statistics Norway. 2011b. *Om statistikken*. Available at: http://www.ssb.no/emner/02/01/10/beftett/. Accessed February, 2011

Syers, K., M. Bekunda, D. Cordell, J. Corman, J. Johnston, A. Rosemarin, and I. Salcedo. 2011. Phosphorus and Food Production. In *UNEP Year Book 2011: Emerging Issues in Our Global Environment,* edited by T. Goverse and S. Bech. United Nations Environment Programme (UNEP) Division of Early Warning and Assessment, Nairobi, Kenya.

Tarkalson, D.D., and R.L. Mikkelsen. 2003. A phosphorus budget of a poultry farm and a dairy farm in the Southeastern U.S. and the potential impacts of diet alterations. *Nutrient Cycling in Agroecosystems* 66:295–303.

Tetra Tech, Inc. 2008. Management of Wapato lake quality. *Project Report for Metro Parks* (http://www.metroparkstacoma.org/files/library/954c2c455c033179.pdf).

Tilman, D., J. Fargione, B. Wolff, C. D'Antonio, A. Dobson, R. Howarth, D. Schindler, W. H. Schlesinger, D. Simberloff, and D. Swackhamer. 2001. Forecasting agriculturally driven global environmental change. *Science* 292:281–284.

Townsend, A. R., and S. Porder. 2011. Boundary issues. *Environmental Research Letters* 6:011001.

Van Drecht, G., A. F. Bouwman, J. Harrison, and J. M. Knoop. 2009. Global nitrogen and phosphate in urban wastewater for the period 1970 to 2050. *Global Biogeochemical Cycles* 23:GB0A03.

Vitousek, P. M., R. Naylor, T. Crews, M. B. David, L. E. Drinkwater, E. Holland, P. J. Johnes, J. Katzenberger, L. A. Martinelli, P. A. Matson, G. Nziguheba, D. Ojima, C. A. Palm, G. P. Robertson, P. A. Sanchez, A. R. Townsend, and F. S. Zhang. 2009. Nutrient imbalances in agricultural development. *Science* 324:1519–1520.

Walker, T. W., and J. K. Syers. 1976. Fate of phosphorus during pedogenesis. *Geoderma* 15:1–19.

Welch, E. B., and J. M. Jacoby. 2001. On determining the principal source of phosphorus causing summer algal blooms in western Washington lakes. *Lake Recovery and Management* 17:55–65.

Whalen, J. K., and C. Chang. 2001. Phosphorus accumulation in cultivated soils from long-term annual applications of cattle feedlot manure. *Journal of Environmental Quality* 30:229–237.

Zhang, H., J. L. Schroder, J. K. Fuhrman, N. T. Basta, D. E. Storm, and M. E. Payton. 2005. Path and multiple regression analyses of phosphorus sorption capacity. *Soil Science Society of America Journal* 69:96–106.

6

Phosphorus Recovery and Reuse

P IS FOR PROCESSING

Hiroko Yoshida, Kimo C. van Dijk, Aleksandra Drizo, Steven W. van Ginkel, Kazuyo Matsubae, Mark Buehrer

Sue Norton-Scott; *Puzzling Out Phosphorus;* Mixed Media, 35" x 45" x 4", 2011

Sue Norton-Scott
Puzzling Out Phosphorus

Scientific Collaborator:
Laura Schreeg, PhD student, Department of Biology, University of Florida

Description of Artwork:
The piece explores two versions of the phosphorus cycle: a "natural" phosphorus cycle, which emphasizes the recycling and the reuse of phosphorus, and an incomplete human-driven cycle where phosphorus is not being recycled. In the natural cycle, phosphorus is released through the decay of plant matter and subsequently reused by plants. In contrast, in the human-driven cycle, phosphorus is mined, applied as fertilizer and then lost from the agricultural system through crop harvesting and runoff/leaching/erosion. Scientist/collaborator Laura Schreeg is pictured in the piece doing research in a Peruvian Amazon jungle.

Viewers can physically move puzzle pieces, which symbolize phosphorus, along a wire from one location to another within the cycles. Incomplete wire connections represent possible solutions for better completing the human-driven cycle in the future, thus illustrating ways to manage the element for a more sustainable future.

About the Artist:
Sue Norton-Scott creates mixed-media art that asks people to think differently about contemporary topics. After graduating from Crane School of Music in Potsdam, New York, the young teacher moved across the country and planted roots in Chandler, Arizona. Now, thirty years later, she is retired from the Mesa Public Schools and lets her creative side flourish. Her surrealistic, three-dimensional art has been displayed in New York City, Chicago, and Sacramento; as well as in the Mesa Contemporary Arts Gallery. Norton-Scott holds a master's degree in Interdisciplinary Humanities and teaches art classes online at Rio Salado College.

BOX 6.1
CHAPTER 6 OBJECTIVES

- Identify the possible P flows that might be intercepted for recovery and reuse.
- Analyze the existing practices for intercepting P from landfills, waterways, and non-arable lands.
- Explore possibilities for retrofitting or reinventing physical infrastructure to enhance interception and recovery of P from human and livestock waste streams.
- Discuss the trade-offs and allied benefits associated with P recovery and reuse projects.

Phosphorus (P) management practices pose significant sustainability challenges. P is supplied to arable land in both mineral and organic forms for agricultural production, but is taken up by plants and animals at limited efficiencies; a considerable amount of P is lost from human society to surrounding environment or deposited in landfills. Prior chapters explore a wide range of options for improving the efficiency of P-use through the food supply chain and for reshaping agricultural and urban landscapes to enhance recycling of P. This chapter focuses on the technological aspects of P recovery and reuse for closing the P cycle at various scales (Figure 6.1).

The efficiency of recycling should be measured by reduction in demand of P fertilizer produced from virgin phosphate ore. Hence, recycling of P is more than just recovering P from the P resources otherwise wasted; it is about redistributing P where and when it is needed without compromising public health and other sustainability imperatives. This requires a chain of physical and institutional infrastructures to connect points of P emissions (supply) with points of agricultural production (demand). Each step incurs capital and operational costs, energy and material inputs, and other environmental emissions, while some strategies for closing the P cycle can circumvent these steps by directly connecting the secondary P sources with the food production system.

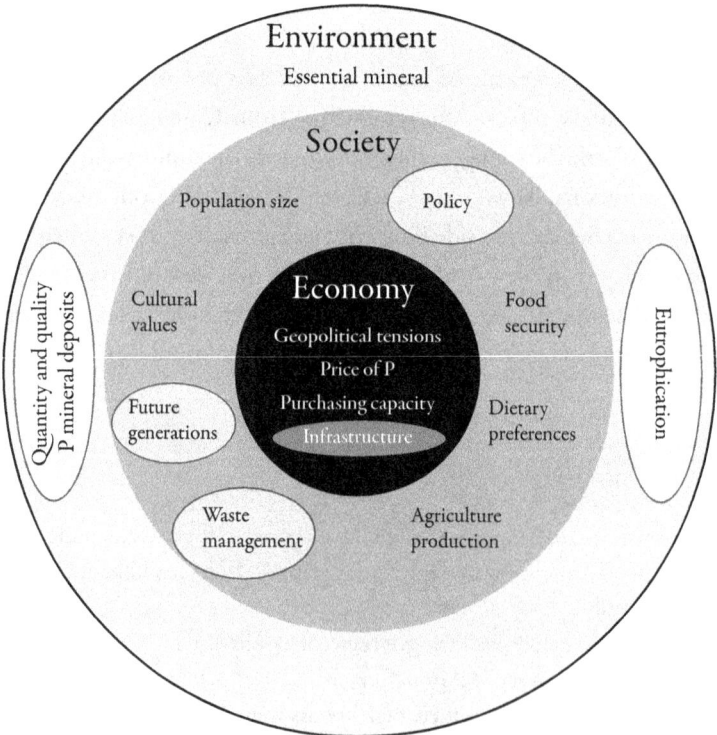

FIGURE 6.1 Sustainability figure for P management in wastewater.

Above all, P recycling plans have to reflect economic, environmental, and social realities. **Figure 6.2a-d** show a summary of P flow analyses for Africa, China, the Netherlands, and Japan. In general, these cases show that agriculture, especially livestock production, represents the largest amount of P flows. P dynamics,

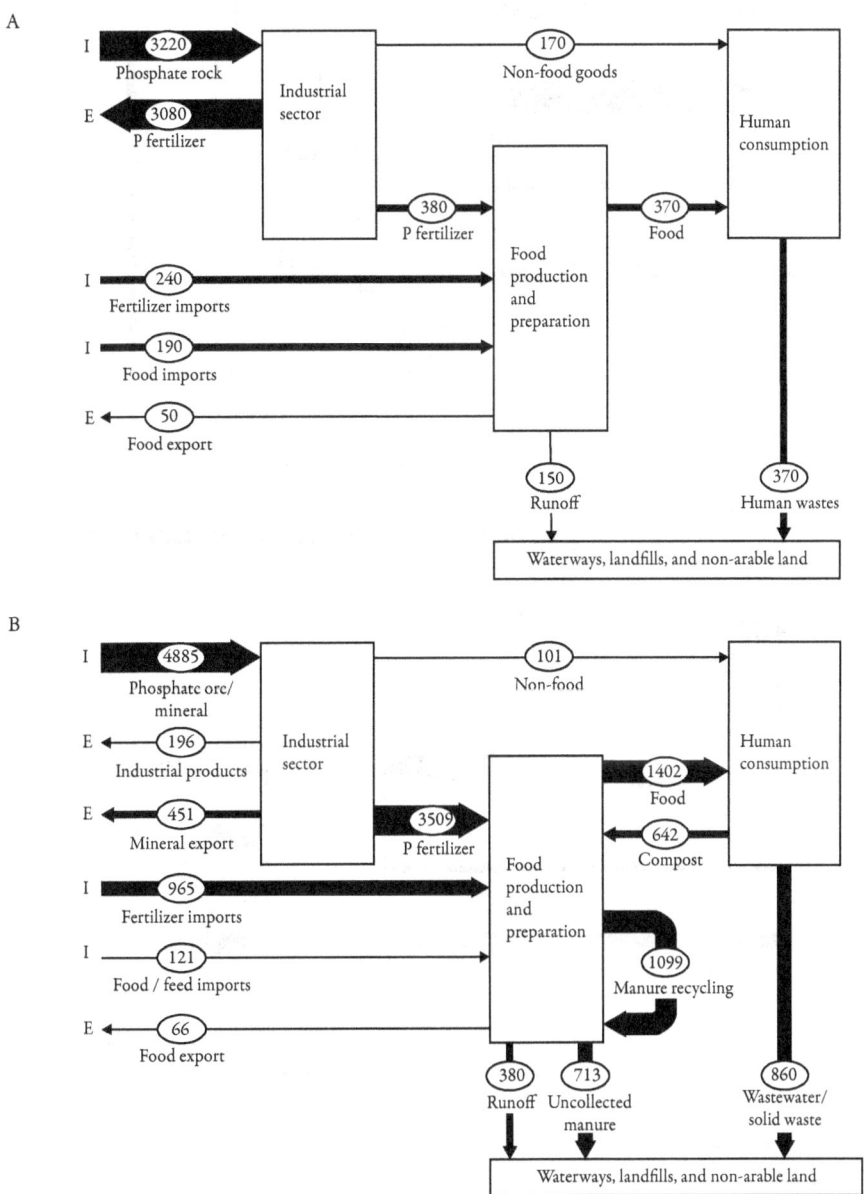

FIGURE 6.2 Comparison of phosphorus flows and stocks (Kt/yr) on the continental (Africa (a)) and national levels (China (b), the Netherlands (c), and Japan(d)). Material Flow Analyses are based on references mentioned in the text. Manure recycling in Africa was not included in the original study, and hence omitted from the figure (a).

116 Phosphorus, Food, and Our Future

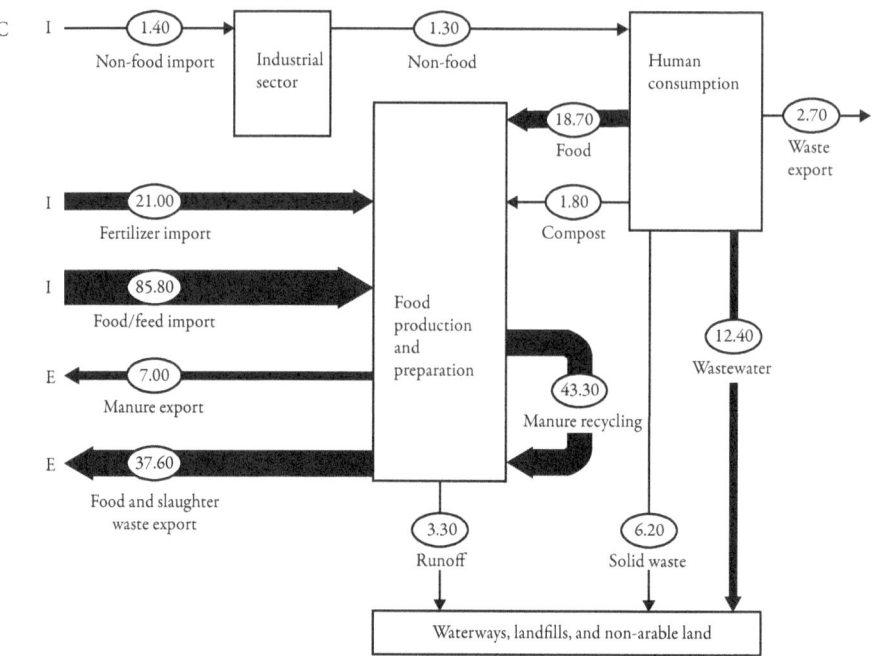

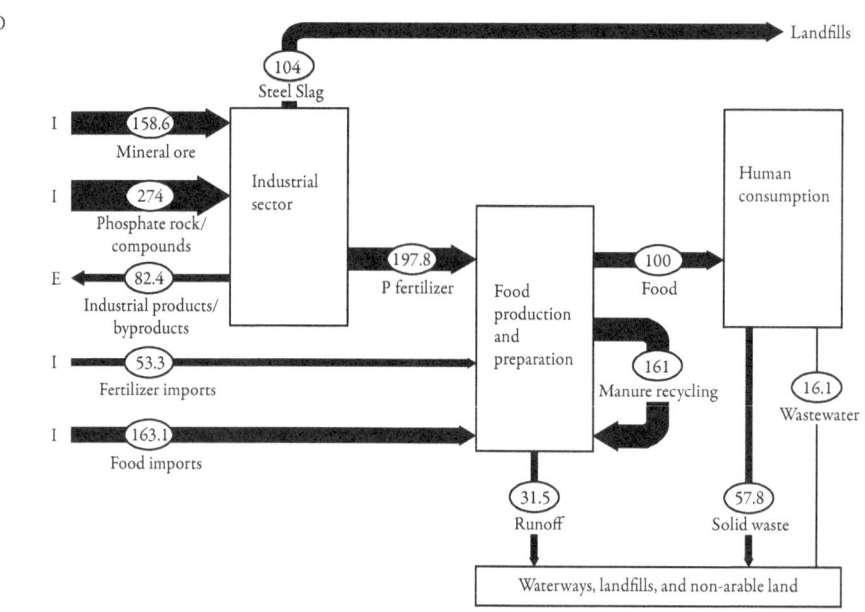

FIGURE 6.2 (*Continued*)

however, vary greatly by region and industrial structure. Africa has low P input into the agriculture sector, although large amounts of P are exported from the continent as mined P-rock from Morocco and the North Africa region. There are still many opportunities to improve yields by easing access to P resources (Cordell et al. 2009). China applies large amounts of mineral P fertilizer to keep up with their ever-increasing demand for food, and there is room to improve phosphorus use efficiency (Liu et al. 2007, Wang et al. 2011). In the Netherlands, due to its high concentration of livestock production, export of P is needed to prevent over-fertilization of cropland ("manure dumping") and subsequent surface water pollution (Schröder et al. 2010). In Japan, due to its high dependency on food imports and high industrial output, industrial byproducts such as steel slag is regarded as one of the promising sources for P recovery (Box 6.2). P in human waste, such as human excreta, organic waste, and agricultural waste, are currently underutilized in all cases.

BOX 6.2

JAPAN: MINING P FROM STEEL-MAKING SLAG

At the global scale, non-agricultural use of P represents less than 10 percent of total P demand (FAO Stat 2011). Yet in Japan, about one-third of total P inputs end up in industrial products/by-products (Matsubae-Yokoyama et al. 2009). Within the industrial sector, P is equally important and non-substitutable as for the food production system. P is a vital ingredient in detergent, pesticides, food and feed additives, and some medicines. P compounds are also used for the etching of semiconductors and other electronic equipment, metal plating, and other metal surface treatment processes. For example, the phosphating process is used to prepare metal surfaces for further painting and deposition of organic agents which prohibit corrosion. Phosphating sludge from metal surface treatment processes contains high concentrations of P (as high as 18 percent P), which can be separated by a steam separation process and recovered in an elemental form used by chemical industries (Matsubae-Yokoyama et al. 2009).

At the national scale, due to its sheer amount of raw mineral input, large amounts of P end up in steel-making slag, which is comparable in size to the total phosphate ore import. P in slag can be further enriched through magnetic separation or acid leaching. Using magnetic separation technologies, up to 65 percent of the P in enriched slag can be recovered and used as a raw material for P fertilizer production, being an example of **industrial symbiosis** (Kubo et al. 2010). Removal of P also reduces the demand of some raw materials and may improve the economic performance of steel mills (Matsubae-Yokoyama et al. 2010).

BOX 6.2 (*Continued*)

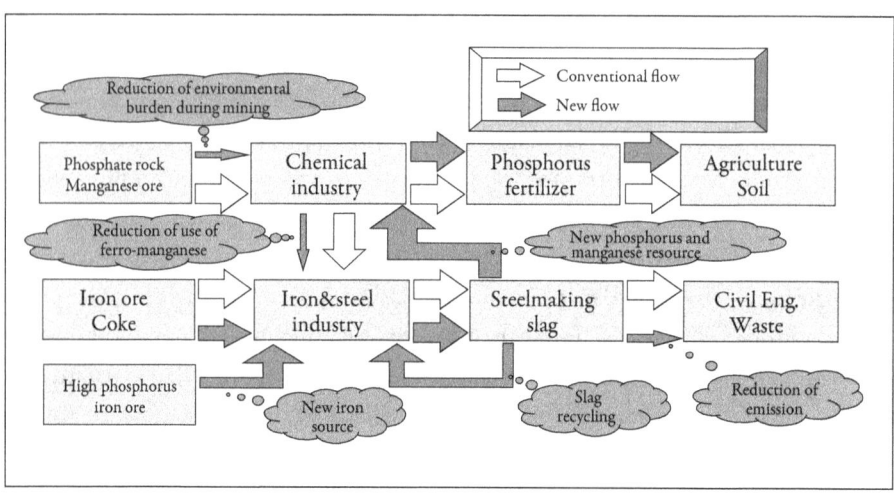

Three outflows of P from human society are covered in this chapter: human wastes (human excreta and organic waste), manure from livestock production, and agricultural and urban runoff water. P in those three flows is largely seen as a pollutant that causes eutrophication. Already some statutory requirements to remove P exist in some industrialized nations. In that case, existing infrastructure could be retrofitted to recover, not only remove, P from the waste streams. For those areas where development of infrastructure is still under way, it is possible to reinvent waste management systems with the intent of nutrient recycling. This chapter showcases some current initiatives and success stories for P recovery and reuse.

RECOVERY OF P FROM HUMAN WASTES

Currently about 14–16 percent of mineral P input into the global food systems ends up in human wastes (Liu et al. 2008, Cordell et al. 2009). Despite the fast expansion of modern sanitation practices, roughly 40 percent of the world population lacks access to **improved sanitation,*** where open defecation is still a common practice (WHO/UNICEF JMP 2008). Households that have a sewer line connection are considered to have access to improved sanitation, but this does not guarantee that the associated wastewater is receiving proper treatment, let alone that P is removed adequately. For example, in South Eastern Europe, over half of collected wastewater is directly discharged, despite the rapid expansion of the wastewater infrastructure in the past decade (EEA 2013). In places already equipped with sewer lines and wastewater treatment facilities, expanding and upgrading these existing infrastructures is a sensible approach, while there is a possibility of moving to alternative less capital and energy-intensive options for areas where sanitary infrastructure is still under development (Figure 6.3).

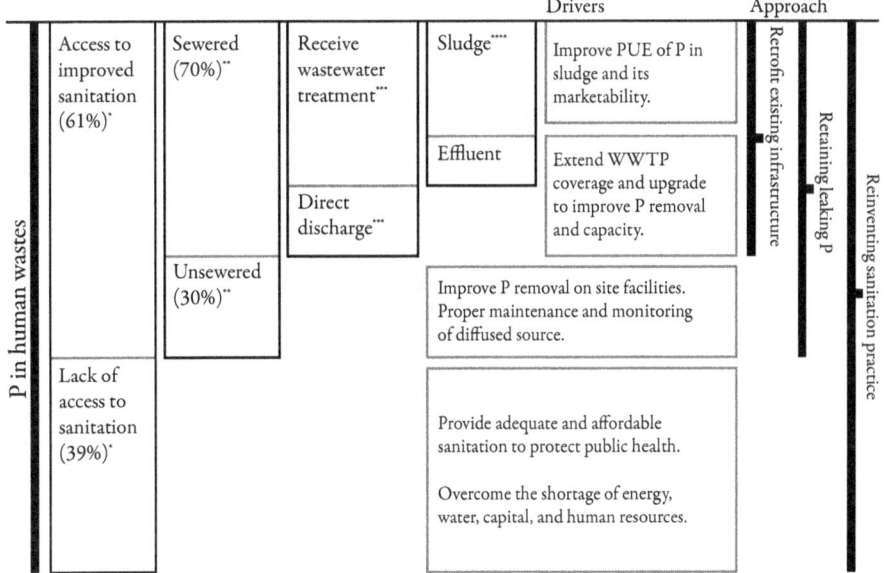

FIGURE 6.3 The state of human waste management and drivers for phosphorus (P) removal or Recovery in 2006.
*WHO/UNICEF JMP; **Average of OECD member states in 2006. The global average could be significantly lower than this but there is not official statistics available at this moment; ***It is estimated that 90% of sewage in developing countries is directly discharged into rivers, lakes and coastal waters without treatment. ; ****The removal efficiency of P from effluent to sludge can vary from 30% for primary treatment, 50% for secondary treatment 90+% for advanced treatment.

Retrofitting Existing Sanitary Infrastructure

Modern urban sanitary infrastructure started with networks of drainage ditches, and pipes, carrying **human excreta*** and other nuisances to nearby water bodies (Schladweiler 2010). Sanitary sewers are often connected with urban runoff (combined sewers) and industrial wastewater with or without pretreatment. The construction of wastewater treatment plants (WWTPs) at the end of these pipelines started in the early twentieth century, though it wasn't until the 1970s that coverage of modern WWTPs was expanded rapidly with the rise of regulations to protect environmental quality in industrialized countries (Balmer and Hultman 1988).

The primary goal of WWTPs is to remove biodegradable organic matter, suspended solids, and pathogens. Nutrients, at some degree, are also removed via generation of sewage sludge (Figure 6.4a). Conventional wastewater treatment plants consist of primary and secondary treatment processes that intercept 30–50 percent of P (Balmer and Hultman 1988, Kristensen et al. 2004). Today, more stringent P removal discharge limit is mandated in industrialized nations, though the extent of this is still limited: 30 percent of the population connected to the public sewers in the United States is only served by a secondary treatment plant without further P removal (OECD 2008).

Both chemical and biological approaches are available to enhance P removal via sludge. Chemical P removal involves the addition of ferric salts, alum, or lime to

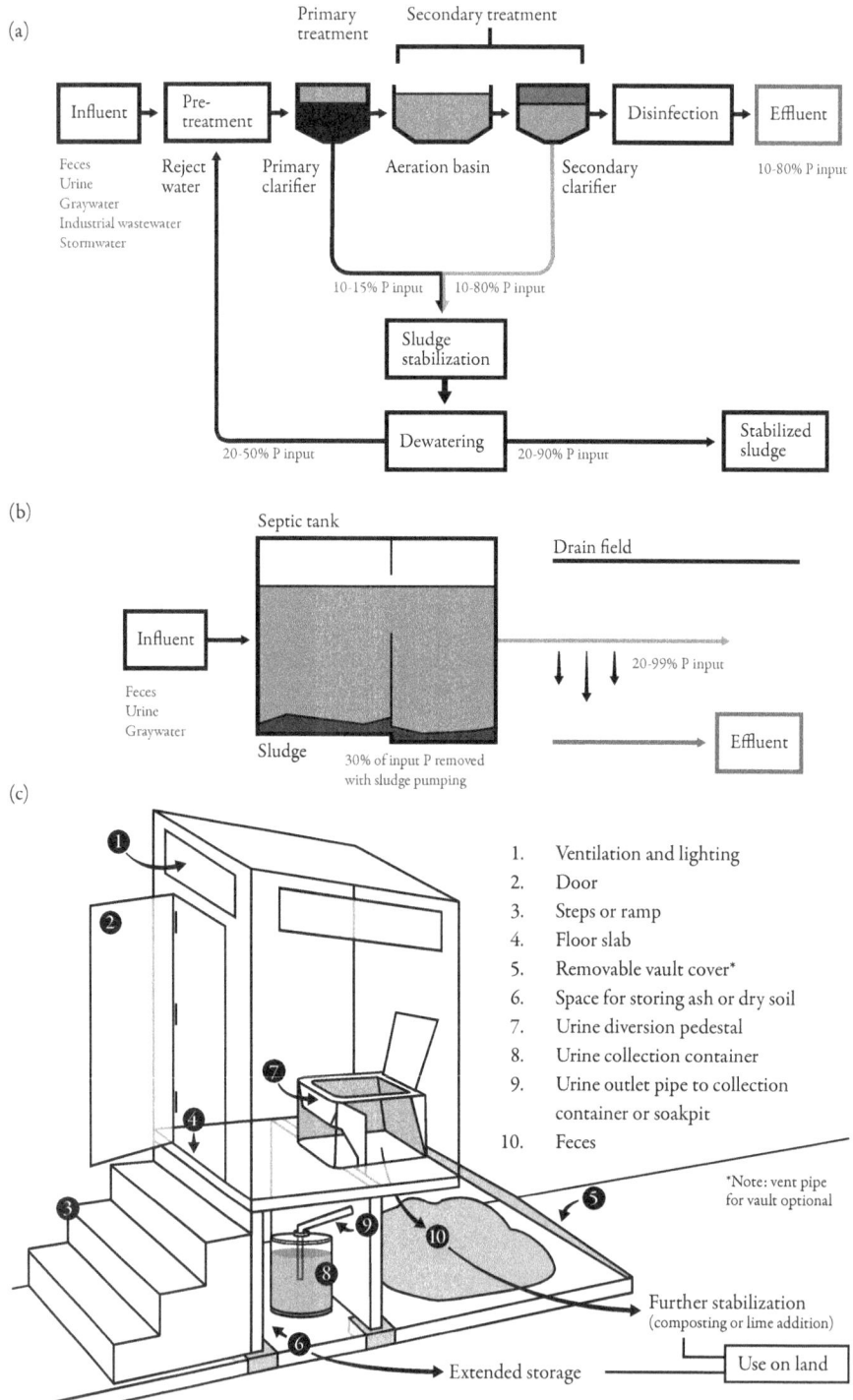

FIGURE 6.4 Example of waste treatment options for P recovery. a) Conventional wastewater treatment plant; b) Septic system; and c) Ecological Sanitation.

wastewater, which in turn creates large quantities of sludge that must be disposed of off-site (Tchobanoglous et al., 2002). P removal from wastewater can be achieved through biological processes such as the Enhanced Biological Phosphorus Removal (EBPR). Temporary exposure to oxygen-depleted conditions triggers certain bacteria present in wastewater to accumulate polyphosphate when they regain access to oxygen. The P-laden biomass is removed from the effluent through succeeding sedimentation processes. A well-run EBPR process can remove >90 percent of P (Kristensen et al. 2004, Tchobanoglous et al. 2002).

Utilization of sludge is the most direct way to recycle P in wastewater. To prevent putrification and reduce pathogen content, sludge is stabilized biologically (composting, anaerobic and aerobic digestion), chemically (lime addition), or thermally (thermal drying, incineration). Some technologies, such as anaerobic digestion, provide the opportunity for energy recovery, which is discussed later in this chapter. Successfully stabilized sludge (biosolids) is also rich in nitrogen and organic matter and could be applied on land as fertilizer or soil conditioner. For instance, in Australia, UK, and France, more than 60 percent of sludge is land applied (Kelessidis and Stasinakis, 2012). However, due to its origin, sewage sludge may contain heavy metals, pathogens, and household chemicals and pharmaceuticals. Hence, land application of sludge has been under vigorous debate (Bengtsson and Tillman 2004, Beecher et al. 2005).

In order to recover P in a more marketable and transportable form, various technologies have been developed and implemented. **Table 6.1** summarizes the major P recovery technologies from the aqueous phase, sludge, and ash. In sludge, P is incorporated with organic matter, while in incineration ash, P is tightly bound to minerals. A part of P accumulated in sewage sludge is released via sludge stabilization processes such as anaerobic digestion, and creates liquor rich in dissolved P (Montag 2008). Remaining P has to be extracted by various technologies such as acid and alkali leaching, hydrolysis, wet ash melting, or thermal treatment. Additional methods, such as advanced oxidation, are needed to convert the non-reactive, organic phosphorus in the sludge to inorganic P (Kato et al. 2007, Sartorius et al. 2011).

Once dissolved, P could be recovered through crystallization or absorption methods. One of the promising P recovery technologies is crystallization of magnesium ammonium phosphate (**struvite*** or MAP) (Jaffer et al. 2002, Sartorius et al. 2011). Addition of magnesium and adjustment of pH induces the precipitation of struvite. Recovered struvite is easy to transport and could be used directly as slow-releasing fertilizer. Struvite is reported to contain much less heavy metals, pathogens, and trace organic pollutants than sewage sludge (Ueno and Fujii 2001, Gell et al. 2011). WWTP operators are already familiar with the term struvite, since it causes scaling

TABLE 6.1

Technology for harvesting P (Kato et al. 2007, Rittmann et al. 2011, and Satorius et al. 2011)

	Process	Description	Form of P recovered
Dissolved P	Struvite (MAP) Crystallization (AirPrex®, Ostara Pearl®, PHOSNIX®, Prisa, and the REM NUT® ion exchange process.)	• Crystallization of PO_4^{3-}, NH_4^+, Mg^{2+} through addition of magnesium and pH control or stripping of carbon dioxide (pH 8.5 or pH 9.8) • Relatively low running cost • Ease of operation in maintaining optimum conditions • Remove nitrogen simultaneously • Not suitable for low concentration P sources	MAP ($MgNH_4PO_4$) Slow-releasing P fertilizer
	HAP Crystallization (Crystalactor®, PRoC)	• Crystallization of PO_4^{3-}, OH^-, Ca^{2+} through Ca compound and high pH (>10) • Generate small amount of by-product • Low running cost • Requires pretreatment (degassing and filtration) • Hard to maintain seeding grain	HAP ($Ca_5(PO_4)_3OH$) Raw material for fertilizer production
	Electrolysis (Fuji Clean CRX)	• Immersing iron electrode into wastewater and passing DC current. Fe iron eluted from the electrode reacts with phosphate and precipitate. • Ease of operation • Simple mechanics allow for application in small-scale/decentralized wastewater treatment systems • Requires maintenance of electrode to prevent the generation of iron hydroxide layer	$FePO_4$
	Adsorption (PROPHOS, RECYPHOS, PHOSIEDI)	• WW is put through a filter tower filled with adsorbents, such as Zr, active alumina or iron-based adsorbents, or beads with surface micropores and an internal network of submicron pores. • Industrial by-products (steel slag, blast furnace slag, coal slag, and iron oxide tailings) can be used • Some adsorbents have very high affinity to P and high removal efficiency • High operational cost and large reactor volume is required • Most applicable to high flows such as WWTP effluents	Various phosphate compounds

P bound to sludge or slurry	Cambi-KREPRO	• Hydrolysis of sewage sludge under pH 1~2 and 150°C and settle out organic compound and heavy metals. P is removed by addition ferric salts from the supernatant • Reduce the sludge volume and save disposal cost • Addition of coagulants and heat for hydrolysis is required	$FePO_4$
	HeatPhos	• Heat sludge from EBPR to 70°C and extract P from the sludge. $CaCl_2$ is added to remove P from the supernatant • Selectively extract the P in sewage sludge • Presence of Fe prohibits the P release	Calcium phosphate
	AquaReci	• Utilize super critical wet oxidation method to convert organics into carbon oxide and water and other innocuous molecules. P is extracted by adding NaOH and lime to the residual sludge. • Dramatically reduce sludge volume in several minutes • High oxygen and energy demand	Calcium phosphate
	Seabourne	• Suitable for recovering nutrients from various biomasses, such as sludge from wastewater treatment plants, manure, and agricultural waste. • Process involves acid leaching, removal of heavy metals, and precipitation of struvite.	Struvite
Incineration Ash and Sludge	Bio-Con	• Add sulfuric acid to the incineration ash to extract P and heavy metals. The extracts are treated through ion-exchange system to recover phosphate, $FeCl_3$, $KHSO_4$ • Reduces the volume of ash • Ion-exchange process is costly	Phosphate
	Sephos & Advanced Sephos	• Uses sulfuric acid and caustic soda • Removes heavy metals.	Sephos—Aluminum phosphates Advanced Sephos—calcium phosphate
	PASH	• Recovers phosphorus as calcium phosphate from incinerated sewage sludge ash and also meat and bone meal ash. • Uses HCl and lime • Removes heavy metals	Calcium phosphate

(Continued)

TABLE 6.1 (*Continued*)

	Process	Description	Form of P recovered
Incineration Ash	ASH DEC	• Involves a chloride dosage and thermal treatment • Removes heavy metals	PhosKraft®
	MEPHREC	• Dried sludge is briquetted with slag-forming substances and coke. The mixture is treated at 2000°C • Removes heavy metals	P_2O_5

and clogging of pipes and poses great operational challenges (Doyle and Parsons, 2002). Purposeful removal of P would not only bring extra income to WWTPs, but also save maintenance costs and ease the operation. P could be precipitated as hydroapatite, which is a form of P present in phosphate ore. Recovered hydroapatite could be supplied directly to the existing fertilizer industry as a raw material (Rittmann et al. 2011, Valsami-Jones 2004).

Reducing P Leaks

Expansion of sewer lines to sparsely populated rural areas is prohibitively expensive and thus decentralized treatment practices are used instead. For example, it is estimated that as much as 26 percent of the population in the United States is connected to on-site wastewater treatment processes (OECD 2008). One of the most common on-site wastewater treatment system (OWTS) processes is the **septic system***, consisting of a septic tank and a drain field (Figure 6.4b). In a septic tank, solids are separated through sedimentation and are then stabilized anaerobically, yet some septic tanks are equipped with an aerator to create an aerobic condition to enhance the removal of pollutants. Effluent is typically conveyed by gravity to a drain field, where soils provide further filtration and effluent polishing, or drain straight to the nearby waterways.

When well-managed, about 30 percent of P inflow is removed by settling of solids in the septic tank. The P removal efficiency of the drain field depends on the P holding capacity of the soil, ranging from 20 to 99 percent initially; however, as the soils become saturated with P, their ability to retain P diminishes over time (Pell et al. 1989, Robertson et al. 1998, Drizo et al. 2002). Even though influent to septic tanks contains P at levels comparable to municipal wastewater, virtually no P regulations exist for septic systems. Sweden is an exception, where implementation of technologies which is expected to remove 70 percent of incoming P is required plants (SEPA 2006).

In addition, due to lack of knowledge and negligence, septic systems are often not properly managed (US EPA 2011a). Septic system failure rates were estimated at the range from 10 to 20 percent, though the same report also warns of the possibility of underestimation due to the difficulty of conducting on-site level monitoring (US EPA 2011). A few studies reported that failed septic systems can contribute as much as 20 percent of the total P input in rural watersheds, directly affecting the quality of receiving waters (Novotny and D'Arcy 1999, May et al. 2010, Withers et al. 2010).

Agricultural and urban stormwater runoff is another diffused P loss to the environment. In the United States, increasing concern has been placed on stormwater runoff, and stormwater **best management practices (BMPs)*** have been studied and implemented to reduce these flows in both agricultural and urban settings. BMPs could be nonstructural (public outreach and educational programs), but could also be the construction of structures to physically reduce the P input to surface waters. The most common structural BMPs for stormwater treatment are detention/retention ponds, exfiltration ponds, bio-retention areas, and buffer strips. Generally, however, P is not recovered from these BMPs (USEPA 2008). Despite large governmental investments over the past 20 years to implement stormwater BMPs, the water quality in the U.S. and other parts of the world continues to deteriorate (WRI 2008).

Constructed wetlands* are low-cost, and low-maintenance alternatives to remove organic matter and suspended solids from various wastewater streams. They are engineered systems that provide treatment of wastewater utilizing physical, chemical, and biological processes that involve soils, wetland vegetation, and the associated microbial communities (Vymazal 2010). Over the past 25 years, constructed wetlands have emerged as a promising solution for treating various types of wastewater effluents, including rural, municipal, industrial (acid mine drainage, landfill leachate), agricultural (dairy, manure runoff, slaughter houses), and urban runoff. This technology is also used for dewatering and stabilizing sewage sludge. It has also been recognized for its high removal efficiency in total suspended solids (85–95 percent), pathogens (85–95 percent), and organic matter (90–98 percent), though the removal efficiencies of N and P remain erratic and rarely exceed 60 and 40 percent, respectively (Moshiri 1993, Kadlec and Knight 1996, Vymazal 2010). Finding ways to improve P removal and recovery remains one of the major scientific challenges among constructed wetlands scientists (IWA 2010).

P can be removed from OWTS by small-scale electrolysis and struvite precipitators (Kato et al. 2007, Ganrot et al. 2009). Yet, such decentralized wastewater processes require technologies that can handle large fluctuations in P concentrations. Research was initiated in the 1990s in the search for novel engineering solutions for reliable and economical P removal from OWTS (Mann 1997, Drizo et al. 1997, Johansson, 1999). Over the course of 10 years, wide ranges of natural and industrial by-products have been tested (Drizo et al. 1999, Johansson-Westholm 2006, Vohla et al. 2011). Of all the investigated material, steel slag aggregates, a recyclable by-product from the steel manufacturing industry, showed the greatest efficiency in removing P from both high- and low-effluent concentrations and made it possible to tap into a largely neglected P source (Drizo et al. 2002, Drizo et al. 2008, Johanson-Whestholm 2006, Shilton et al. 2006, Vohla et al. 2011). OWTS can be used either as standalone treatment systems or can be incorporated with BMPs (Drizo et al. 2008).

One of the commercially available decentralized P removal system is PhosphoReduc™. PhosphoReduc™ treatment systems consist of one or multiple filter units filled with iron- and/or calcium-based filtration material graded to particular sizes. P is removed from wastewater through sequestration within the filtration material at specific hydraulic residence times. The system can be placed in drainage pathways of agricultural lands and urban landscapes, removing more than 90 percent of both dissolved and particulate P, suspended solids, and bacteria (AUTM 2011). Once the filter units have reached their usable life, P and minerals retained by the filtration material can be reused as a fertilizer. Researchers in Sweden have shown that blast furnace steel slag could be used as a soil amendment in agriculture, while electric arc furnace steel slag produced in mini mills has potential as a slow-release P fertilizer (Johansson and Hylander 1998, Hylander 2006, Bird and Drizo 2009).

Reinventing the Sanitary Infrastructure for P Recycling

Despite the advancements in the past century, 2.6 billion people still lack access to adequate sanitation and diarrheal disease remains the second-leading cause of death among children under five (WHO/UNICEF JMP 2008). As the world population is growing rapidly, especially in urban areas, there is an urgent need to expand the coverage of the sanitation infrastructure to prevent the transmissions of waterborne disease and water pollution.

Conventional WWTPs require large capital investment, well-trained staff, and a steady supply of energy and chemicals to maintain its function. All of these are in short supply in many parts of the world. Water is another limitation factor. The current flush-and-discharge scheme demands large amounts of water: one individual on average produces 50 L of feces and 400–500 L of urine annually, which are flushed away with 9,000–27,000 L of freshwater (Werner et al. 2004, USEPA 2011b). As half of the human population is expected to face water shortages by 2030, these methods of human waste disposal might not be feasible in the future (Lüthi et al. 2011).

Ecological sanitation,* or **EcoSan** for short, has been developed specifically to overcome these problems, while achieving the recovery of nutrients in human excreta (Werner, Mang et al. 2004; Langergraber and Muellegger 2005). EcoSan technologies are generally waterless processes; human excreta are collected through gravity or vacuum sewerage. Pathogen destruction is achieved by dehydration, pH adjustment (addition of ash, urea, or lime to increase pH), or a composting process. In all cases, the treated human excreta are stored for 6–24 months to ensure its safe use on farmland (Figure 6.4c).

Source separation and the appropriate treatment for each fraction are the core principles of the EcoSan process. Urine and feces are separated via separate outlets in the toilet. Unlike feces, urine contains far less pathogenic organisms and could be safely applied on land after simply stored for a prolonged time. Urine diversion reduces water content in the feces and thus enhances the treatability of human feces. Moreover, 50–60 percent of P in human excreta is concentrated in urine in a soluble plant-available form (Kirchmann and Pettersson 1995).

EcoSan technologies require a minimum input of energy and chemicals. The technology is simple enough to be managed by nonexperts and facilities can be built with locally available materials. A non-governmental organization, Sustainable Organic Integrated Livelihood (SOIL), is based in Haiti, where a legacy of natural resource mismanagement runs deep. Another wave of devastation hit Haiti in 2010 with the magnitude 7.0 earthquakes, which put extra pressure on a country that was already suffering from having the lowest level of access to sanitation in the world. Outbreaks of commutable diseases soon followed. Since the earthquake, SOIL has installed more than 200 toilets at camps for internally displaced personnels, providing safe, clean, and dignified sanitation options (*The New York Times* 2010). There are already many success stories like SOIL across the world, which can be found at The EcoSanRes (www.ecosanres.org) and The Sustainable Sanitation Alliance (www.susana.org).

The very concept of ecological sanitation could be realized in a built environment as well. Urine-diversion technologies have been adapted at the city scale in Europe, Australia, and China (Lienert and Larsen 2010). It has also been adopted by state-of-the-art **green building*** projects on the building level (Box 6.3).

BOX 6.3
DESIGNING A LIVING BUILDING

The Living Building Challenge (LBC) is a green building certification program that promotes the most advanced measurement of sustainability in the built environment. In order to obtain LBC certification, a building has to achieve 20 Imperatives.

For instance, the Net Zero Water Imperative requires that "100% of the water use must come from captured precipitation or closed loop water systems" and the Urban Agriculture Imperative requires that "all projects must integrate opportunities for urban agriculture appropriate to the scale and density. A of the project." A living building must treat and manage all wastewater (i.e., graywater and blackwater) on-site. This is achieved by various on-site treatment options such as composting toilets, constructed wetlands, and other available wastewater treatment technologies. The closed-loop water and nutrient design of a living building makes it possible to capture and recycle the majority of P, as well as other nutrients, that are produced by the occupants in the building and at the same time provides opportunities for on-site food production.

As a project example to meet the criteria of closed-loop design for water and nutrients including P, the figure on page 129 shows the concept design of a commercial office building called the Bullitt Center, "living" in the middle of Seattle, Washington. Rainwater is harvested off the roof and stored in a cistern to meet the overall water demand. The rainwater is then filtered and disinfected with UV light before being pumped to fixtures. Urine is harvested from the waterless urinals and urine-diversion toilets which is stabilized and stored for fertilizer use. Graywater from the sinks, showers, and toilets is used to dilute the nutrient-rich stabilized urine at a ratio of approximately 8:1 prior for applications in the greenhouse enclosure, vegetated planter boxes, or constructed wetlands. Solid wastes from the toilets (composted with wood chips) are to be stored and treated in composting units prior to application as a solid fertilizer. Organic food scraps from the building can be composted in an on-site vermicomposting system. The stabilized urine and compost products are all used as fertilizer for the agricultural applications to produce various types of vegetables.

BOX 6.3 (*Continued*)

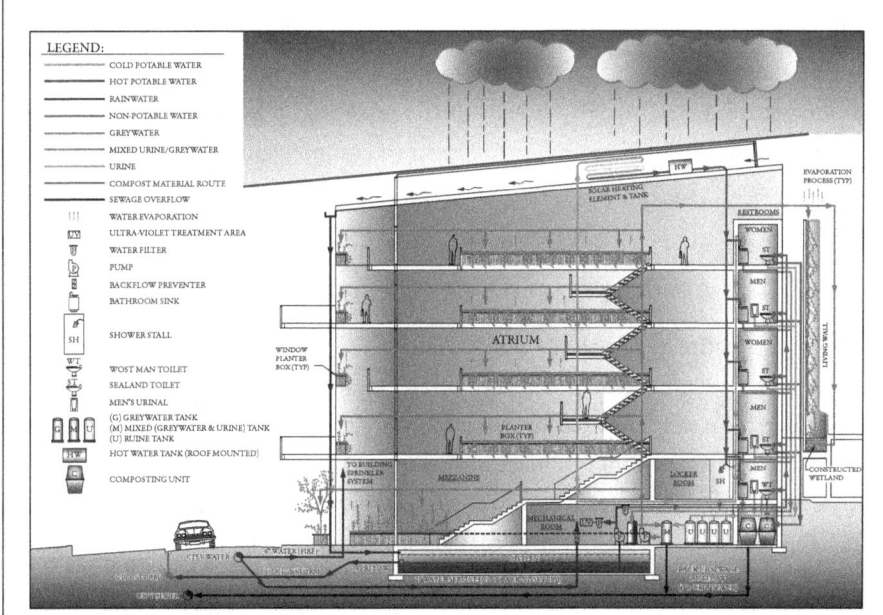

Photo Credit:
2020 ENGINEERING;Wellspring Building; Bellingham, WA (360) 671-2020, ext. 103

Capturing the P from the Solid Waste Stream

The organic fraction of municipal solid waste (OFMSW) is another major P flux, containing up to 26 percent (Switzerland) and 31 percent (the Netherlands) of P leaving a household (Schröder et al. 2010). It is often commingled with other types of waste. Thus, the first step of P recovery from OFMSW is separation. Once segregated, animal feeding of the food wastes is the most direct way of bringing P in OFMSW back to the food production system. This has been done historically, and USEPA/USDA's Food Waste Recovery Hierarchy encourages animal feeding of food waste after pasteurization over other treatment methods (USEPA 2011c).

OFMSW can be stabilized biologically and chemically prior to land application just like sewage sludge. These processes prevent vector and pest attraction and reduce odor problems. The most common practice is composting, which can be done in composting bins in backyards or at a centralized location. Compared to composting, anaerobic digestion and pyrolysis have the advantages of additional energy recovery. The diversion of organic waste from conventional waste streams in recent years has shown promise in reducing greenhouse gas emissions, by preventing the emissions of methane from landfills (European Environmental Agency 2011). Diversion of organic waste, including sewage sludge, has been encouraged under the EU Landfill Directive in 1999, yet, to date, only 30 percent of the potentially available biodegradable waste is separately collected and processed to compost in all EU member states (Delgado, et al. 2009).

RECOVERY OF P FROM LIVESTOCK MANURE

Globally, it is estimated that 36–40 percent of P input into the food production systems goes through livestock and ends up in manure (Liu et al. 2008, Cordell et al. 2009). Livestock can be raised on pastures, rangelands, and forested lands. The P cycle in the pasture system is almost closed: 2–5 percent leaves the system as meat and/or milk and 75–90 percent are cycled back to the grazing field (Betteridge et al. 1986). P is lost from the system via erosion, but manure could be directly deposited to waterways when grazing livestock has access to them. Redistribution of manure over the feeding ground can be achieved by adjusting forage and grazing intensity, animal density, or changing the location of the drinking water source or shade (Dao et al. 2010).

Yet, as described in chapter 5, over the past several decades livestock production has become more concentrated and intensified across the world. Still, mixed livestock–crop production is the dominant form of livestock production, but often P is brought to farm as concentrated feed or mineral P supplement, resulting in on-farm and regional imbalances of nutrients. Temporal imbalance of supply and demand is also an issue; while manure is produced year round, application of fertilizers is permitted only at specific times of the year. Surplus of manure, in both geographical and temporal terms, not only means inefficient utilization of nutrients in manure, but can also result in manure spills and dumping, and subsequent water-quality degradation (FAO, 2006).

Manure is managed as dry solid manure or liquid slurries. Manure is often removed and stored in open pits or piles. Liquid manure and wastewaters are sent to detention ponds or lagoons for settling out the solids fraction and reducing the volume through evaporation. Lagoons also serve as a temporary storage facility before land application. P could seep through the lagoon bottom due to failure or nonexistence of liners in these facilities. P can be lost from manure management systems through runoff water from animal housing, surfaces of animal pens, manure storage facilities, and/or leaching from manure-amended fields (Dao et al. 2010).

P loss can be reduced by construction of adequate storage facilities and the precise application of manure based on crop nutrient requirements. Addition of alum or ferric salts to liquid manure could reduce P runoff (Kleinman et al. 2002). Adoption of erosion control measures and buffer strips along waterways and other BMPs offers additional lines of defense. More recently, there has been a move to recognize manure not only as a source of nutrients and water pollution but also as a source of renewable energy, providing yet another driver of change for P recycling (Box 6.4). Treatment of manure and improvement of nutrient management practices could bring regional P budgets more in balance (Kara et al. 2012).

BOX 6.4
ADDING ENERGY INTO THE EQUATION

Manure is the second-largest sustainable source of biomass from U.S. agricultural lands, comprising 18 percent of the total supply (Perlack et al. 2005). Bio-energy projects provide an opportunity for concentrating the flow of P-laden manure to one location and granting economy of scale. There are two approaches to convert biomass into bio-energy: biochemical and thermochemical conversion.

Anaerobic digestion* is one of the most commonly used biochemical conversion processes. It can be done by installing an impermeable cover over existing lagoons, or constructing reactor tanks for higher performance. Germany is by far leading the way with more than 4000 on-farm anaerobic digesters in operation (Weiland 2009). Anaerobic digestion is a process of microbial degradation of complex organic matter under oxygen-deprived conditions. Biogas, predominantly methane, is produced, which can be combusted for heat and electricity production, or upgraded to natural gas quality (Cantrell et al. 2008). The heat produced by electricity production can be used to dry final products for improved transportability.

Besides renewable energy production, anaerobic digestion provides a wide range of benefits at the farm, local, and societal level. Besides manure, a wide range of organic substrates, such as agricultural residues, organic waste, and sewage sludge-can be treated by anaerobic digestion for renewable energy production. From the P recovery standpoint, anaerobic digester provides an additional storage capacity and mitigate temporal imbalance of P supply and demand. After digestion, roughly 25 percent of P is partitioned to the liquid phase, while the rest remains in the solid fraction. The liquid fraction can be applied directly on land as high-quality fertilizer, rich in readily available dissolved N and P for plants. Dissolved P can be extracted through the technologies listed in Table 6.1 (Cantrell et al. 2008, Ritmann

(Continued)

BOX 6.4 *(Continued)*

et al. 2011). The solid fraction can be sold as a bedding material, fertilizer, or fuel (Schoumans et al. 2011).

Thermochemical conversion is a high-temperature chemical process, in which a chemical reforming process breaks apart the bonds of organic matter and reforms these intermediates into char, synthesis gas, and highly oxygenated bio-oil. Pyrolysis gasification and direct liquefaction are some examples. These could be combined with algae production for extra P removal and biomass production (Cantrell et al. 2008).

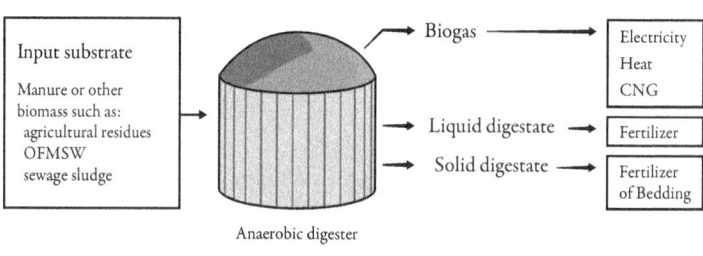

Conceptual representation of anaerobic digestion process.

Besides livestock manure, runoff from barnyards, feedbunks, and crop field can also carry P into the surrounding environment. For instance, subsurface tile drains are used to improve drainage on agricultural fields, and they also open a pathway for large amounts of P to be transported to surrounding waters with minimum intervention (Milburn and MacLeod 1991; Sharpley et al. 2000; Kinley et al. 2007). More recently, Drizo and Twohig (2011) found that even without application of manure, storm runoff water in tile drains contains P far exceeding the eutrophication limits. Currently, however, there are no management practices that can effectively reduce diffuse P pollution from tile drainage, and appropriate technologies are being sought.

Bones and other wastes from the slaughtering process could be another P flow lost in the animal production. Animal carcasses normally contain a minimum of 3.5–5 percent total phosphorus (Miles and Jacob 2011). More than 4 billion kg of protein meal are produced by the U.S. rendering industry each year, and most of that is meat and bone meal (Coxworth 2011). In the Netherlands, slaughter waste accounted for 23 percent of total P exports in 2005. Two-thirds of this P resides in processed bone meal and chips, and about one-third is used for non-agricultural purposes such as porcelain production (Smit et al. 2010). Since the bovine spongiform encephalopathy (BSE) crisis, in Europe, deceased animals and slaughter waste of bovines are incinerated to prevent BSE from spreading (Schröder et al. 2010).

CHALLENGES AND FUTURE PERSPECTIVE

The opportunities to close the P cycle exist at various scales—within the building, the neighborhood, or even at the national or regional scale; and at various technology levels—from simple composting, to retrofitting conventional wastewater treatment plants, stormwater BMPs, to state-of-the-art green buildings and renewable energy projects. We have already seen some of the success stories, and many technologies are to be developed and implemented. However, challenges still lie ahead in establishing closed P cycles and appropriate environmental and economic impact assessments.

Balancing Costs and Benefits

Every decision is accompanied by cost and benefits. Some P recovery strategies require large capital investments, and at the time of writing, it is almost impossible to justify projects solely from the P recycle perspective. Molinos-Senante et al. (2011) conducted the economic analysis of installation of a struvite precipitator in major Spanish WWTPs. Annualized capital, operational, and maintenance cost is $19.3/kg P/year recovered, while the internal benefits for the utility operation is $0.51/kg P/year for sales of struvite. However, P recovery projects come with a wide range of allied benefits imperative to other sustainability agendas, such as provisioning of adequate and affordable sanitation, preventing surface water pollution, and the production of renewable energy. It also directly and indirectly saves the operation and maintenance cost of existing facilities. The same study suggested that the plant would benefit $2.17/kg P/year for reduction of chemical use and parts replacement, and another $37.1/kg P/year for prevention of eutrophication and conservation of non-renewable resource (Molinos-Senante et al. 2011).

Given the extent of eutrophication worldwide (WRI 2008, US 2010), P regulations will be revisited in the next 5–10 years, requiring the retrofitting of existing WWTPs to lower P discharge levels to below 1.0 mg/L (Presidential Chesapeake Bay Executive Order 2009, US EPA 2010). P discharge limits of 1.0 mg/L incurs 14.9–37.3 \$/kg of P recovered for the mid-size city (38,000 m^3/day) (Jiang et al.

2005). This might also be a sunk cost, driving the implementation of P recovery projects.

Cost and benefits also come in environmental terms as well. Additional energy and material inputs are needed to implement the projects, while recovery of P could reduce the demand of mineral P, thus reducing the environmental burden associated with it. For instance, application of manure or stabilized sludge results in sequestration of carbon to soil and improves water-holding capacity and fertility. More holistic measure of evaluating environmental performance, such as Life Cycle Assessment, could be adopted to facilitate an informed decision-making process (Lundie et al. 2004, Palme et al. 2005).

Reuse, Not Dispose

The goal of P recovery from waste streams is to reduce the dependence on primary P sources and to provide P resource to populations that otherwise have limited access to P fertilizers. Cooperation and involvement of both supply and demand sides of P-use value chain is vital for the success of projects. Relative effectiveness of the processing of P from human and livestock wastes to mineral P varies greatly and depends on the composition of waste and treatment measures (Elliott and O'Connor 2007). Uncertainty in fertilization potential of recovered P is another risk to farmers and is a potential hurdle for P reuse. For example, struvite is a slow-releasing P fertilizer and some farmers are still somewhat skeptical about its fertilizer value (Schröder et al. 2010). Mismatches between crop fertility requirements and nutrient content of recycled P products are another hurdle for incorporating recovered P into a nutrient management plan. More research on the fertilizer value of recovered P and the development of a support program, like crop insurance, would encourage the use of recovered P and reduce dependence on mineral P fertilizer.

It is evident that human waste and livestock manure contain pathogens and other forms of pollutants. The creation of a nutrient recycling loop could potentially open a pathway for pollutant exposure and pathogen transmission (Carballa et al. 2004, Busetti et al. 2005, Burkholder et al. 2007). The risk can be minimized through education and outreach programs for information sharing through the supply chain. Then, in between, the P recovery process must be operated, monitored, and maintained properly. Proper institutional arrangements should be established to ensure proper operation and fiscal stability.

Rather than the top-down, supply-based approach traditionally taken by infrastructure projects, a P recycle plan should be formulated based on the **communicative planning approach,*** and harvest agreement and cooperation among the stakeholders through P supply chain (Lüthi et al. 2011). Educating a new generation of engineers, planners, architects, and scientists who can manage multi-objective projects and work as liaisons between institutions and stakeholders with conflicting values is essential for advancing P recycling across the world (Ozawa and Seltzer 1999).

Evolving with Regulation

Just as physical infrastructures have developed over time, the regulatory and institutional framework also evolve as new social needs emerge. However, the changes may come slowly. Diffused sources such as effluent from OWTS, urban stormwater, and agricultural runoff water represent a significant portion of P outflow from human society. Despite its detrimental effect on the water bodies, the regulations on intercepting this P flow are sparsely applied. In some cases, regulation could be a setback, rather than a driver of change. For instance, in many parts of industrialized nations, building codes and health standards do not allow on-site wastewater treatment in the urban setting, including the use of gray water (water from sinks, bathtubs, and washing machines) for irrigating lawns. This could be a hurdle for implementing close-nutrient systems in green buildings (Cascadia Green Building Council 2011).

However, regulations are put into place to protect the public through setting guidelines and providing resources to address the pending issues. Establishment of statutory requirements calls for financial investments necessary for fostering innovative technologies: research and development, bench-scale testing to full-scale demonstrations, monitoring and permitting, and implementation and commercialization of technologies (Shilton et al. 2006). These incentives create a new market for entrepreneurs to explore (Stark 2005). In Europe, the EU Urban Wastewater Directive set a requirement for P removal in 1991 (91/271/EEC). Following the implementation of this regulation, a few European countries (Denmark, Finland, Netherlands, Norway, and Sweden) increased the coverage of nutrient removal at WWTPs, and N and P discharge decreased by 60 and 30 percent over 10 years (Kristensen et al. 2004). In some countries like Sweden, Germany, and Japan, strategic recovery of P has been included in the national initiatives.

The following chapters will discuss more about overcoming the above-mentioned challenges, including personal beliefs and value systems for P recovery and reuse (chapter 7), capacity and partnership building among stakeholders (chapter 8), and a framework for making future changes (chapter 9).

BOX 6.5
CHAPTER 6 SUMMARY

- There are significant P recycling opportunities for recovering P from human excreta, agricultural and industrial waste, and runoff waters.
- Technologies and strategies are already available to recycle P from above-mentioned sources at the various scales and technological levels.
- These technologies could improve transportability and marketability of recovered P and contribute in creating a closed P cycle where P is reused for food production.
- Appropriate strategies for P recycling should be identified in consideration of institutional, societal, and economic conditions.
- P recycling by recovery and reuse of P could include allied benefits of other sustainability imperatives: provision of adequate and affordable sanitation, preventing water pollution, renewable energy production, and direct and indirect improvement of the economic performance of waste management practices.

REFERENCES

AUTM, the Association of University Technology Managers. 2011. Academic Filtration Innovations Aim to Solve What Ails a Perishable Resource: Water. Published in the 2011 Edition of a Better World Report. Respond, Recover, Restructure: Technologies Helping the World in the Face of Adversity, pp. 78–83. www.betterworldrpoject.net

Balmer, P., and B. Hultman. 1988. Control of phosphorus discharge; present situation and trends. *Hydrobiologia* 170: 305–319.

Beecher, N., E. Harrison, N. Goldstein, M. McDaniel, P. Field, and L. Susskind. 2005. Risk perception, risk communication and stakeholder involvement for biosolids management and research. *Journal of Environmental Quality* 34: 122–128.

Bengtsson, M., and A. M. Tillman. 2004. Actors and interpretation in an environmental controversy: the Swedish debates on sewage sludge use in agriculture. *Resource Conservation & Recycling* 42: 65–82.

Betteridge, K., W. G. K. Andrews, and J. K. Sedcole. 1986. Intake and excretion of nitrogen, potassium and phosphorus by grazing steers. *Journal of Agricultural Science* 106:393–404.

Bird, S.C., and A. Drizo. 2009. Investigation on phosphorus recovery and reuse as soil amendment from electric arc furnace slag filters. *Journal of Environmental Science and Health, Part A: Toxic/Hazardous Substance and Environmental Engineering* 44 (13): 1476–1483.

Burkholder, J., B. Libra, P. Weyer, S. Heathcote, D. Kolpin, P. S. Thorne, and M. Wichman. 2007. Impact of waste from concentrated animal feeding operations on water quality. *Environmental Health Perspectives* 115(2): 308–312.

Busetti, F., S. Badoer, M. Cuomo, B. Rubino, and P. Traverso. 2005. Occurrence and removal of potentially toxic metals and heavy metals in the wastewater treatment plant of fusina (Venice, Italy). *Industrial & Engineering Chemistry Research* 44(24): 9264–9272.

Cantrell, K. B., T. Ducey, K. S. Ro, and P. G. Hunt. 2008. Livestock waste-to-energy generation opportunities. *Bioresource Technology* 99:7941–7953.

Carballa, M., F. Omil, J.M. Lema, M. Llompart, C. García-Jares, and I. Rodríguez. 2004. Behavior of pharmaceuticals, cosmetics and hormones in a sewage treatment plant. *Water Research* 38(12):2918–2926.

Cascadia Green Building Council. 2011. Regulatory Pathways to Net Zero Water: Guidance for Innovative Water Projects in Seattle. http://cascadiagbc.org/resources/Regulatory PathwaystoNetZeroWater.pdf

Chesapeake Bay Executive Order. 2010. http://executiveorder.chesapeakebay.net/default.aspx

Cordell, D., J. O. Drangert, and S. White. 2009. The story of phosphorus: Global food security and food for thought. *Global Environmental Change* 19: 292–305.

Coxworth, B. 2011. Partially-biodegradable plastic made from waste bone meal. Retrieved from http://www.gizmag.com/bioplastic-Dao

Dao, T.H. and R.C. Schwartz. 2010. Effects of manure management on phosphorus biotransformations and losses during animal production. In Bünemann, E. K., Oberson, A., and Frossard, E. 2010. *Phosphorus in Action,* 407–428. Springer-Verlag, Berlin, Heidelberg, Germany.

Delgado, L., A. S. Catarino, et al. 2009. *End-of-Waste Criteria: Final Report.* Luxembourg, European Commission, Joint Research Centre.

Doyle, J. D., and S. A. Parsons. 2002. Struvite formation, control and recovery. *Water Research* 36: 3925–3940.

Drizo, A., J. Cummings, D. Weber, E. Twohig, G. Druschel, and B. Bourke. 2008. New Evidence for Rejuvenation of Phosphorus Retention Capacity in EAF Steel Slag. *Environmental Science and Technology,* 42: 6191–6197.

Drizo, A., C. Forget, R. P. Chapuis, and Y. Comeau. 2002. Phosphorus removal by EAF steel slag—A parameter for the estimation of the longevity of constructed wetland systems. *Environmental Science and Technology* 36:4642–4648.

Drizo, A., A. C. Frost, K. A. Smith, and J. Grace. 1997. The use of constructed wetlands in phosphate and ammonium removal from wastewater. *Water Science and Technology* 35 (5):95–102.

Drizo, A., A. C. Frost, K. A. Smith, and J. Grace. 1999. Physico-chemical screening of phosphate-removing substrates for use in constructed wetland systems. *Water Research* 33 (17):3595–3602.

Drizo, A., and E. Twohig. 2011. Phosphorus—a major water pollutant and a crucial natural resource in peril. *1st World Sustainability Forum.* http://www.wsforum.org

Elliott, H. A., and G. A. O'Connor. 2007. Phosphorus management for sustainable biosolids recycling in the United States. *Soil Biology and Biochemistry* 39: 1318–1327.

European Environment Agency. 2013. Urban waste water treatment. Retrieved from http://www.eea.europa.eu/data-and-maps/indicators/urban-waste-water-treatment/ urban-waste-water-treatment-assessment-3

European Environment Agency. 2011. *Waste opportunities: Past and future climate benefits from better municipal waste management in Europe.* Copenhagen, Denmark, European Environment Agency.

Food and Agriculture Organization. 2011. *FAOSTAT.* Rome: Statistics Division, FAO.

Food and Agricultural Organization 2006. Livestock's Long Shadow: Environmental Issues and Options. Retrieved from ftp://ftp.fao.org/docrep/fao/010/a0701e/A0701E00.pdf

Ganrot, Z., J. Broberg, and S. Byden. 2009. Energy efficient nutrient recovery from household wastewater using struvite precipitation and zeolite adsorption techniques: a pilot study in

Sweden. In *International Conference on Nutrient Recovery from Wastewater Streams,* edited by K. Ashley, D. Mavinic, and F. Koch. *Vancouver, 2009.* IWA publishing, London, UK.

Gell, K., F. J. de Ruijter, P. Kuntke, M. de Graaff, and A. L. Smit. 2011. Safety and effectiveness of struvite from black water and urine a phosphorus fertilizer. *Journal of Agricultural Science* 3(3):67–80.

Hylander, L. D., A. Kietlinska, G. Renman, and G. Siman. 2006. Phosphorus retention in filter materials for wastewater treatment and its subsequent suitability for plant production. *Bioresource Technology* 97:914–921.

IWA 2010. Proceedings of the 12th International Conference on Wetland Systems for Water Pollution Control, held in Venice (Italy), October 4–8, 2010.

Jaffer, Y., T. A. Clark, P. Pearce, and S. A. Parsons. 2002. Potential phosphorus recovery by struvite formation. *Water Research* 36:1834–1842.

Jiang, F., M. B. Beck, R. G. Cummings, K. Rowles, and D. Russell. 2005. Estimation of Costs of Phosphorus Removal in Wastewater Treatment Facilities: Adaptation of Existing Facilities. Water Policy Working Paper #2005-011.

Johansson, L. 1999. Blast furnace slag as phosphorus sorbents—Column studies. *Science of Total Environment* 229: 89–97.

Johansson, L., and L. Hylander. 1998. Phosphorus removal from waste water by filter media: Retention and estimated plant availability of sorbed phosphorus. *Zeszyty Problemowe Postı¨p´ow Nauk Rolniczych* 458: 397–409.

Johansson-Westholm, L. 2006. Substrates for phosphorus removal—Potential benefits for on-site WW treatment? *Water Research* 40: 23–36.

Kadlec R. H., and R. L. Knight. 1996. *Treatment Wetlands.* Lewis Publishers, Chelsea, MI.

Kara, E. L., C. Heimerl, T. Killpack, M. C. Von de Bogert, H. Yoshida, and S. R. Carpenter. 2012. Assessing a decade of phosphorus management in the Lake Mendota, Wisconsin watershed and scenarios for enhanced phosphorus management. *Aquatic Science* 74(2):241–253.

Kato, F., M. Takaoka, K. Oshita, and N. Takeda. 2007. Present state of phosphorus recovery from wastewater treatment. *JSCE Magazine of Civil Engineering* 63(4):413–423.

Kelessidis, A., and A. S. Stasinakis. 2012. Comparative study of the methods used for treatment and final disposal of sewage sludge in European countries. *Waste Management* 32(6):1186–1195.

Kinley, R. D., R. J. Gordon, G. W. Stratton, G. T. Patterson, and J. Hoyle. 2007. Phosphorus Losses through Agricultural Tile Drainage in Nova Scotia. *Canadian Journal Environment Quality* 36:469–477.

Kirchmann, H. and S. Pettersson. 1995. Human urine: Chemical composition and fertilizer use efficiency. *Fertilizer Research* 40: 149–154.

Kleinman, P. J. A., A. N. Sharpley, B. G. Mover, and G. F. Flwinger. 2002. Effect of mineral and manure phosphorus sources on runoff phosphorus. *Journal of Environmental Quality* 31(6):2026–33.

Kristensen, P., S. Nixon, and B. Fribourg-Blanc. 2004. Outlook on Nutrient Discharge in Europe from Urban Waste Water Treatment Plants. *European Environmental Agency Topic Center on Water.*

Kubo, H., K. Matsubae-Yokoyama, and T. Nagasaka. 2010. Magnetic Separation of Phosphorus Enriched Phase from Multiphase Dephosphorization Slag. *ISIJ International* 50(1):59–64.

Langergraber, G., and E. Muellegger. 2005. Ecological Sanitation—a way to solve global sanitation problems? *Environment International* 31(3): 433–444.

Lienert, J., and T. A. Larsen. 2010. High acceptance of urine source separation in seven European countries: a review. *Environmental Science Technologies* 44 (2):556–566.

Liu, Y., J. Chen, A. P. J. Mol, and R. U. Ayres. 2007. Comparative analysis of phosphorus use within national and local economies in China. *Resources Conservation and Recycling* 51:454–474.

Liu, Y., G. Villalba, R. U. Ayres, and H. Schroder. 2008. Global phosphorus flows and environmental impacts from a consumption perspective. *Journal of Industrial Ecology* 12(2): 229–247.

LeBlanc, R. J., P. Matthews, and R. P. Richards. 2008. *Global atlas of excreta, wastewater sludge, and biosolids management: welcome uses of a global resource.* United Nations Human Settlements.

Lundie, S., G. M. Peters, and P. Beavis. 2004. Life cycle assessment for sustainable metropolitan water systems planning. *Environmental Science & Technology* 38(13): 3465–3473.

Lüthi, C., A. Panesar, T. Schütze, A. Norström, J. McConville, J. Parkinson, D. Saywell, and R. Ingle. 2011. Sustainable Sanitation in Cities: A Framework for Action. Sustainable Sanitation Alliance. Retrieved from http://www.eawag.ch/forschung/sandec/publikationen/sesp/dl/sustainable_san.pdf

Mann, R. A. 1997. Phosphorus adsorption and desorption characteristics of constructed wetland gravels and steelworks by-products. *Australian Journal of Soil Resources* 35: 375–384.

Matsubae-Yokoyama, K., H. Kubo, and T. Nagasaka. 2010. Recycling effects of residual slag after magnetic separation for phosphorus recovery from hot metal dephosphorization slag. *ISIJ International* 50 (1):65–70.

Matsubae-Yokoyama, K., H. Kubo, K. Nakajima, and T. Nagasaka. 2009. Material flow analysis of phosphorus in Japan: considering the iron and steel industry as a major sector. *Journal of Industrial Ecology* 13(5): 687–705.

May, L., C. Place, M. O. O'Malley, and B. Spears. 2010. The impact of phosphorus inputs from small discharges on designated freshwater sites. NERC/Centre for Ecology & Hydrology, Retrieved from http://nora.nerc.ac.uk/9242/.

Milburn, P., and J. MacLeod. 1991. Considerations for tile drainage water quality studies in temperate regions. *Applied Engineered Agriculture* 7:209–215.

Miles, R.D., and J. P. Jacob. 2011. Using Meat and Bone Meal in Poultry Diets. Retrieved from http://edis.ifas.ufl.edu/ps024

Molinos-Senante, M., F. Hernández-Sancho, R. Sala-Garrido, and M. Garrido-Baserba. 2011. Economic feasibility study for phosphorus recovery processes. *AMBIO* 40 (4): 408–416.

Montag, D. 2008. *Phosphorus recovery in wastewater treatment: Development of a procedure for integration into municipal wastewater treatment plants.* (Phosphorrückgewinnung bei der Abwasserreinigung—Entwicklung eines Verfahrens zur Integration in kommunale Kläranlagen.) Retrieved from http://darwin.bth.rwth-aachen.de/opus3/volltexte/2008/2298/pdf/Montag_David.pdf

Moshiri, G.A. (ed.). 1993. *Constructed Wetlands for Water Quality Improvement.* Lewis Publishers, Boca Raton, FL.

Novotny, V., and B. D'Arcy (eds.). 1999. *Diffuse Pollution '98. Proceedings of the IAWQ 3rd International Conference on Diffuse Pollution, held in Edinburgh*, UK, August 31–September 4, 1998. IWA Publishing, London, UK.

Organization for Economic Co-Operation and Development. 2008. OECT Statistics on Environment: Wastewater Treatment. http://stats.oecd.org/Index.aspx?DataSetCode=WATER_TREAT

Ozawa, C. P., and E. P. Seltzer. 1999. Taking our bearings: Mapping a relationship among planning practice, theory and education. *Journal of Planning Education and Research* 18: 257–266.

Palme, U., M. Lundin, A. M. Tillman, and S. Molander, S. 2005. Sustainable development indicators for wastewater systems—researchers and indicator users in a co-operative case study. *Resources Conservation and Recycling* 43(3): 293–311.

Pell, M., and J. Nyberg, J. 1989. Infiltration of wastewater in a newly stored pilot sand filter system:1. *Reduction of organic matter and phosphorus. Journal Environmental Quality* 18:451–457.

Perlack, R. D., L. L. Wright, A. F. Turhallow, R. L. Gramham, B. J. Stokes, and D. C. Erbach, 2005. Biomass as a Feedstock for a Bioenergy and Bioproducts Industry: The Technical Feasibility of Billion-Ton Annual Supply. http://feedstockreview.ornl.gov/pdf/billion_ton_vision.pdf

Pütz, P. 2008. *Elimination and determination of phosphates. Practice Report Laboratory Analysis & Process Analysis Nutrients Phosphate*. Hatch Lange Ltd DOC040.52.10011, September 2008.

Rittmann, B. E., B. Mayer, P. Westerhoff, and M. Edwards. 2011. *Capturing the lost phosphorus*. Chemosphere. 84(6): 846–853. Robertson, W., S. Schiff, and C. Ptoceck. 1998. Review of phosphorus mobility and persistence in 10 septic system plumes. *Ground Water* 36(6):1000–1010.

Sartorius, C., J. van Horn, and F. Tettenborn. 2011. *Phosphorus Recovery from Wastewater—State of the art and future potential*. Proceeding from International Conference on Nutrient Recovery and Management held in Miami, FL, January 9–12 2011.

Schladweiler, J. C. 2010. Tracking down the roots of our sanitary sewers. http://www.sewerhistory.org/

Schoumans, O. F., W. H. Rulkens, O. Onema, and P. A. I. Ehlert. 2011. Phosphorus Recovery from Animal Manure: Technical Opportunities and Agro-Economical Perspective. Wageningen, Plant Research International, Wageningen University and Research Centre.

Schröder, J. J., D. Cordell, A. L. Smit, and A. Rosemarin. 2010. *Sustainable Use of Phosphorus* (European Union tender project ENV.B.1/ETU/2009/0025). Wageningen, The Netherlands, Plant Research International, Wageningen University and Research Centre: 122.

Swedish Environmental Protection Agency. 2006. Naturvårdsverkets allmänna råd [till 2 och 26 kap. miljöbalken och 12–14 och 19 §§ förordningen (1998:899) om miljöfarlig verksamhet och hälsoskydd] om små avloppsanordningar för hushållsspillvatten (NFS 2006:7) (in Swedish). Swedish Environmental Protection Agency, Stockholm, Sweden.

Sharpley, A. N., B. Foy, and P. Withers. 2000. Practical and innovative measures for the control of agricultural phosphorus losses to water: An overview. *Journal of Environmental Quality* 29:1–10.

Shilton, A. N., I. Elmetri, A. Drizo, S. Pratt, R. G. Haverkamp, and S. C. Bilby. 2006. Phosphorus Removal by an "Active" Slag Filter—a Decade of Full Scale Experience. *Water Research* 40 (1): 113–118.

Smit A. L., J. C. Curth-van Middelkoop, W. van Dijk, H. van Reuler, A. J. de Buck, P. A. C. M. van de Sanden. 2010. *A Quantification of Phosphorus Flows in the Netherlands through Agricultural Production, Industrial Processing and Households*. Wageningen, Plant Research International, Wageningen University and Research Centre.

Stark, K. 2005. Phosphorus Recovery—Experiences from European countries. Retrieved from http://www2.lwr.kth.se/forskningsprojekt/Polishproject/rep12/StarkSthlm19.pdf

Tchobanoglous, G., F. L. Burton, and H. D. Stensel. 2002. *Wastewater Engineering: Treatment and Reuse*. McGraw-Hill, New York.

The New York Times. 2010. Haiti, Nearly a Year After. Retrieved from http://www.nytimes.com/2010/12/02/opinion/02kristof.html?scp=5&sq=%22Sasha%20Kramer%22&st=cse

Ueno, Y., and M. Fujii. 2001. Three year experience in operating and selling recovered struvite from full scale plant. *Environmental Technology* 22: 1373–1381.

United Nations Population Fund. 2010. State of the World Population. Retrieved from http://www.unfpa.org/public/swp2010

United States Environmental Protection Agency. 2008. National Menu of Stormwater Best Management Practices. Retrieved from http://cfpub.epa.gov/npdes/stormwater/menuofbmps/index.cfm

United States Environmental Protection Agency. 2010. Impaired Waters and Total Maximum Daily Loads. U.S. Environmental Protection Agency. Retrieved from http://www.epa.gov/owow/tmdl/, 2010.

United States Environmental Protection Agency. 2011a. EPA600/R-00/008 Chapter 1: Background and Use of Onsite Wastewater Treatment Systems.

United States Environmental Protection Agency. 2011b. Water Sense: Toilet FAQ. Retrieved from http://www.bewatersmart.org/RebatePrograms/WaterSenseHigEfficencyToilets/ToiletsFAQ.html

United States Environmental Protection Agency. 2011c. Food Waste Recovery Hierarchy. Retrieved from http://www.epa.gov/osw/conserve/materials/organics/food/fd—gener.htm#food-hier

Valsami-Jones, E. 2004. *Phosphorus in Environmental Technology: Principles and Applications*. IWA Publishing, London, UK.

Vohla, C., M. Kõiv, H. J. Bavor, F. Chazarencc, and Ü. Mander. 2011. Filter materials for phosphorus removal from wastewater in treatment wetlands—A review. *Ecological Engineering* 37(1): 70–89.

Vymazal, J. 2010. Constructed Wetlands for Wastewater Treatment—A Review. Water 2010, 2, 530–549; doi: 10.3390/w2030530. http://www.mdpi.com/2073-4441/2/3/530/pdf

Wang, F., J. T. Sims, L. Ma, Z. Dou, and F. Zhang. 2011. The phosphorus foot print of China's food chain: Implication for food security, natural resource management and environmental quality. *Journal of Environmental Quality* 40: 1081–1089.

Weiland, P. 2009. *Status of biogas upgrading in Germany, IEA bioenergy task 37 workshop on biogas upgrading*. Tulln, Austria.

Werner, C., H. P. Mang. 2004. *Ecosan—Introduction of closed-loop approaches in wastewater management and sanitation—A supra-regional GTZ-project*. Springer-Verlag, Berlin.

Withers, P. J. A., H. P. Jarvie, and C. Stoate. 2010. Quantifying the impact of septic tank systems on eutrophication risk in rural headwaters. *Environmental International* 37(3):644–653.

World Health Organization/United Nations Children's Fund Joint Monitoring Programme for Water Supply and Sanitation. 2008. http://www.wssinfo.org/about-the-jmp/introduction/

World Resources Institute. 2008. World Hypoxic and Eutrophic Coastal Areas. http://www.wri.org/map/world-hypoxic-and-eutrophic-coastal-areas

7

Cultural Beliefs, Values, and the Biogeochemical Cycling of P

P IS FOR PEOPLE

Timothy Crews, James Cotner, Carol McCreary

Randy Zucker
The Latest Poop about P

Scientific Collaborator:
Dan Childers, Professor, School of Sustainability, Arizona State University

Description of Artwork:
Our "Toilet Seat Art" piece addresses phosphorus sustainability and food security from the context of the human phosphorus cycle. We use decoupage techniques and mixed media to connect scientific information with related photographic and design elements using a "canvas" that represents the circular nature of the human phosphorus cycle while also being a well-known (and practical!) part of that cycle (i.e., human waste). A scientific review of the human phosphorus cycle as it relates to phosphorus sustainability challenges, which will be published in the February issue of *Bioscience* (Childers et al. 2011), forms the scientific framework for our piece. We use actual copy from this manuscript as a "text as art" background in our decoupage. Layered over this background, several key figures from the manuscript as well as excerpts and key phrases from the text are featured as images, as are photographs depicting the key points. Our presentation of the material and images is organized into four "chapters" (or sides of the seat): (Side 1–top of lid) Background on the importance of

phosphorus and where it comes from (i.e., the mining of mineral phosphorus); (Side 2–inside of lid) The importance of phosphorus to our food production and the inefficiencies of its use by the "Agri-Industrial Complex"; (Side 3–top of seat) The problems with [what is now] a very "leaky" and wasteful human phosphorus cycle; and (Side 4–bottom of seat) Sustainable solutions to the phosphorus challenge, which include low-impact food production and recycling, to name just two. A faint wash of yellow provides a unifying visual effect and represents pure elemental phosphorus (which is yellow in color) as well as phosphorus-rich animal excretions. Our piece is both decorative and fully functional—we demonstrate the latter through our installation. Notably, our installation base (the toilet) is recycled (from Stardust Industries in Mesa, a local business that specializes in recycling building materials and supplies). The "toilet paper" roll (a familiar accompaniment to our piece and necessary addition to our installation) features labels from different fertilizer bags, symbolic of this link between geologic phosphorus and our production of food (and the associated waste of phosphorus in its production and consumption).

About the Artist:
Randy Zucker is a self-taught "Outsider" artist currently working and showing in Phoenix, Arizona. She is best known for her "Digital Transformation" Images and Decoupage Toilet Seats. She is co-manager of the MonOrchid Artist Collective, past Treasurer of Artlink Phoenix, and represents her collective to the Roosevelt Row Merchant Association. Randy is currently spearheading a year-long Community Art Project which will, through a blog and other communications, involve the public in the conception and creation of a piece of Public Art made from pennies donated by art aficionados of all shapes, sizes, and backgrounds.

Randy Zucker; *The latest poop about P*; Decoupage, 18" x 14" (seat), 2010

Cultural Beliefs and the Cycling of P 145

BOX 7.1
CHAPTER 7 OBJECTIVES

- Explore how cultural values, beliefs, and traditions that do not explicitly pertain to phosphorus can nevertheless profoundly influence the fluxes of P in the biosphere.
- Describe how individual and societal perspectives on genetic engineering can translate into policies that may influence the phosphorus use efficiencies of crops.
- Evaluate how religious expectations or restrictions regarding diet help determine overall fertilizer demand in crop production.
- Assess the implications of cultural taboos or acceptance of using human excreta as fertilizer in agricultural production on P sustainability.

INTRODUCTION

Throughout this volume, natural and social scientists attempt to reconcile the complex **biogeochemistry*** of phosphorus (P) with equally complex societal demands for this vital element. Efforts made to move this reconciliation forward often take the form of policy recommendations that are based on scientific evaluations of P stocks, and the efficiency of their uses in agriculture. Yet, much of what drives the human–P relationship has less to do with on-the-farm P assimilation efficiency (PAE), and more to do with complex cultural values that are unrelated to P per se (Figure 7.1).

The collective behavior of 7 billion people has important consequences for biogeochemical cycles. We examine several ways in which P and human values are intertwined. Although humanity consists of a multitude of cultures, the globalization of communication and culture presents both an opportunity and a challenge from a biogeochemical perspective. The opportunity is that human choices and behaviors can be very rapidly influenced at the global scale, but the challenge is how to open up discussions of options and opportunities to foster more sustainable behaviors or norms. While the emphasis here is on cultural values and choices that relate to P biogeochemistry, many of the factors that contribute to wiser usage of P also should facilitate more efficient use of most of our limiting resources, such as N, water, and energy.

The extent to which food systems maintain relatively closed P cycles, in which P molecules are reused in production and consumption linkages, is at the heart of food security and social sustainability. We present a range of examples that span multiple scales and cultures to illustrate how P fluxes vary as a function of societal values. Examples include: 1) choices to use or not use genetically engineered crops;

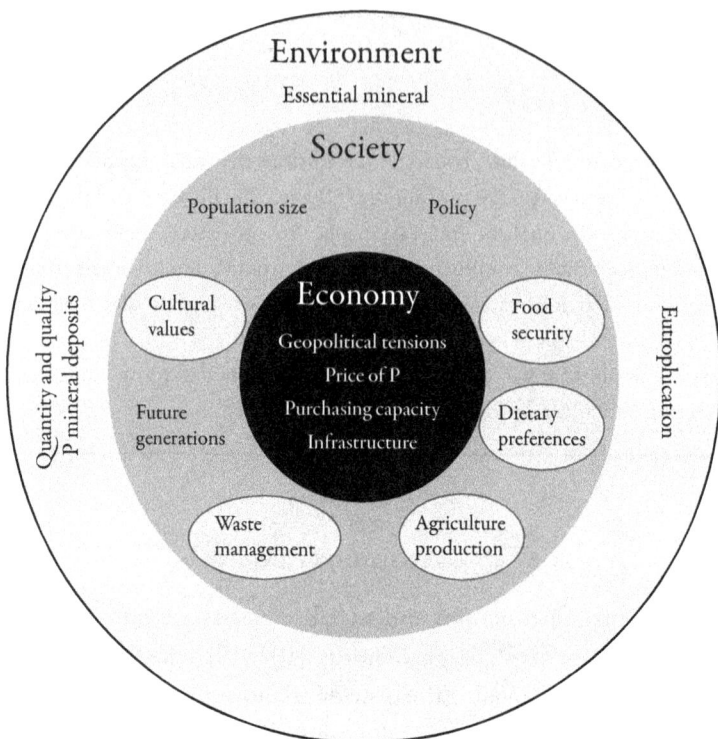

FIGURE 7.1 Social issues in relation to sustainability thinking. The image illustrates the three large considerations of sustainability as a nested model where environment encompasses society, and economy as part of society.

2) choices of consuming animal vs. vegetal products; and 3) choices to use human excreta to fertilize agricultural fields, or dispose of it as waste.

CULTURAL VALUES AND FOOD PRODUCTION TECHNOLOGIES: THE EXAMPLE OF GENETIC ENGINEERING

The contentious and complex social debates around the development and deployment of transgenic crops reflect a diverse range of human values informed by ecological, economic, social, and ultimately moral considerations. The term genetic engineering encompasses a range of gene manipulation technologies and methodologies with applications for basic research (mapping genomes), human health (gene therapy and production of useful proteins), and development of transgenic organisms (moving specific genes from one organism to another) (Nicholl 2008). Transgenic breeding allows for genes to be moved between taxonomically unrelated organisms (e.g., fish and plants). In contrast, traditional breeding methods are constrained to working with closely related species.

The biological, economic, and ethical considerations that bear on the choice to develop and deploy genetically engineered (GE) crops are many. Some argue that food security is profoundly compromised when farmers adopt GE crops, particularly when farmers from less-developed countries become reliant on seeds owned by corporations in more developed countries (e.g., Shiva 2000). In contrast, others contend that GE crops hold great promise to benefit the world's poorest farmers, and efforts to obstruct their adoption robs the most economically vulnerable of valuable strategies for improving their lives (Paarlberg 2009).

There are similarly divergent views on how GE crops might impact the environment. Many of the **transgenic crops*** developed to date are open-pollinated, and thus produce viable seed. Thus, there is concern associated with the escape of ecologically relevant traits—such as N fixation or even improved P uptake—from transgenic crops into wild populations (Stewart et al. 2003). Others contend that the benefits of GE crop adoption to ecosystem health far outweigh potential ecological hazards (Edwards et al. 2009). This debate extends to the effects of the consumption of GE crops on human health and to religious questions regarding humanity's role in designing biological systems (Drees 2011).

Just as the positions of individuals and groups within nations vary tremendously regarding the planting and consumption of GE crops, so do the policy positions between nations. The three most populated countries, China, India, and the United States, have policies that generally support the planting of GE crops, and thus these nations rank 6, 4, and 1, respectively, in area planted to GE crops (Figure 7.2) (ISAAA 2010). The European Union has been the most influential force in resisting adoption of genetically modified crops and foods (Paarlberg 2009). At present, only Monsanto's MON810 Bt corn variety has been approved for use in the European Union, although six individual countries, including France and Germany, have banned its production within their borders. Strongly influenced by Europe's position on GE crops, all but four countries in Africa (South Africa, Kenya, Burkina Faso, and Egypt) currently ban the production and in some cases importation of GE crops. However, Tanzania, Uganda, Malawi, Mali, Zimbabwe, Nigeria, and Ghana all are reported to be conducting field trials, which may lead to a lifting of the GE ban in these nations (Kumwenda 2011). In contrast to China, Japan has strong resistance to the consumption and production of GE crops, while policies from other East Asian countries are more ambiguous. The GE policy stances of countries in Latin America are also very diverse. Peru recently invoked a 10-year ban on the production of GE crops, while Bolivia appears poised to reverse its previous tepid position on GE crops, by increasing production of engineered crops beyond a limited acreage of soybeans (*Sydney Morning Herald* 2011).

Biotech Crop Countries and Mega-Countries*, 2010

*17 biotech mega-countries growing 50,000 hectacres or more of biotech crops.

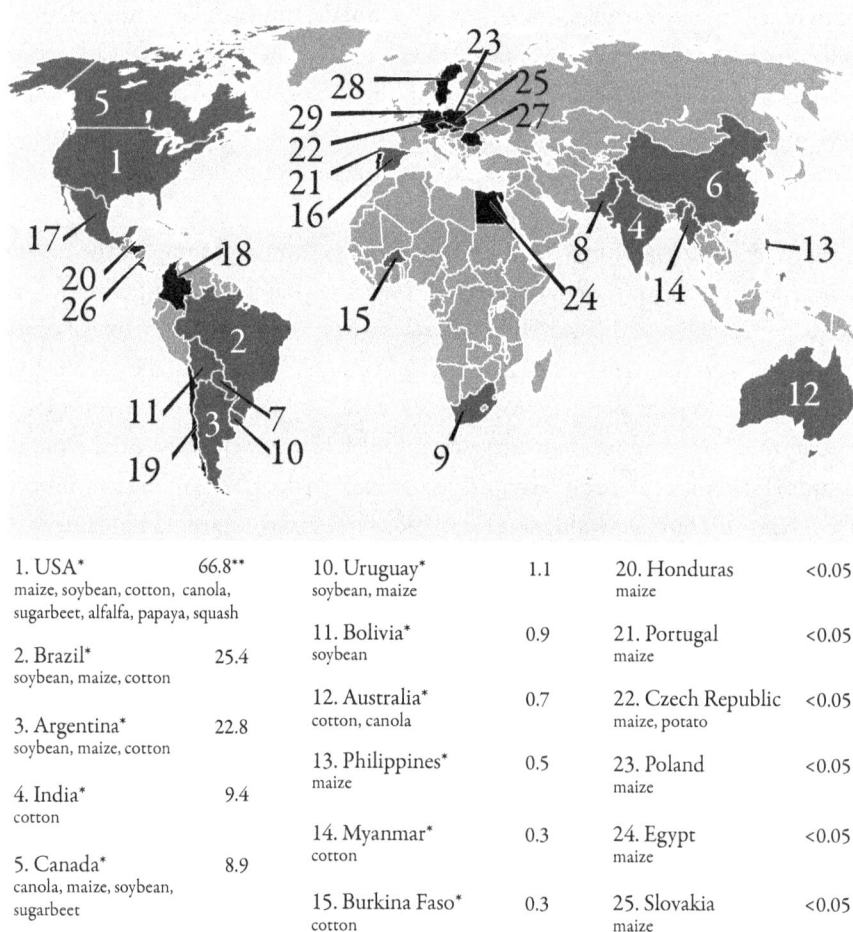

1. USA* maize, soybean, cotton, canola, sugarbeet, alfalfa, papaya, squash	66.8**	10. Uruguay* soybean, maize	1.1	20. Honduras maize	<0.05
2. Brazil* soybean, maize, cotton	25.4	11. Bolivia* soybean	0.9	21. Portugal maize	<0.05
3. Argentina* soybean, maize, cotton	22.8	12. Australia* cotton, canola	0.7	22. Czech Republic maize, potato	<0.05
4. India* cotton	9.4	13. Philippines* maize	0.5	23. Poland maize	<0.05
5. Canada* canola, maize, soybean, sugarbeet	8.9	14. Myanmar* cotton	0.3	24. Egypt maize	<0.05
6. China* cotton, tomato, poplar, papaya, sweet pepper	3.5	15. Burkina Faso* cotton	0.3	25. Slovakia maize	<0.05
		16. Spain* maize	0.1	26. Costa Rica cotton	<0.05
7. Paraguay* soybean	2.6	17. Mexico* cotton, soybean	0.1	27. Romania maize	<0.05
8. Pakistan* cotton	2.4	18. Colombia cotton	<0.05	28. Sweden potato	<0.05
9. South Africa* maize, soybean, cotton	2.2	19. Chile maize, soybean, canola	<0.05	29. Germany potato	<0.05

**Amount of biotech crops shown by million hectacres.

FIGURE 7.2 Genetically engineered crops planted in different countries in 2010 (ISAAA 2010). The table depicts country's rank of total land planted to GE crops.

SUSTAINABLE PHOSPHORUS AND GENETIC ENGINEERING

On greater than 95 percent of land planted to GE crop varieties, crops function either to resist insect herbivory or tolerate herbicides. Under development, however, are an increasing number of crops featuring other agronomic attributes. One area of research that many contend has the potential to help improve the sustainability of agriculture is the use of genetic engineering to improve uptake or use efficiency of inorganic P (P_i) in soils by crops (Ramaekers et al. 2010). Interest in improving P-use efficiencies in agriculture stems from the two overarching P sustainability concerns discussed elsewhere in this volume: rapidly diminishing reserves of affordable P-bearing ores and eutrophication of water bodies caused by fertilizer runoff from agriculture (chapters 2, 3 and 5; Cordell et al. 2009).

In order to understand GE P-uptake, it is important to clearly define "available" P as compared to "total P." Soils range broadly in their total P content, depending on their original parent material and degree of pedogenic development. Young soils that have not been long exposed to weathering at the Earth's crust tend to have greater total P content than soils that have been exposed to precipitation and high temperatures for thousands to millions of years (Crews et al. 1995). While total P content is a useful indicator of the ultimate P resources a soil might contain for agricultural exploitation, it is not a very good predictor of P availability to the biota (see chapter 4). In alkaline soils, phosphate bonds with calcium, magnesium, and carbonates, and is remarkably insoluble and unavailable to plants and microbes (Sharpley et al. 1997). So a soil may be very high in total-P, but have exceedingly low concentrations of P readily available to crops. Similarly, under acid soil conditions, phosphate forms strong, high-energy bonds with iron and aluminum oxides, which are prevalent in clays of more highly weathered soils. In general, the ratio of available P: total-P is highest in slightly acid (pH 6.5–6.8) soil conditions.

When inorganic P fertilizers are applied to soils that have not been consistently fertilized in the past, a large percentage of the applied P becomes tightly bound or "fixed" to calcium or oxides of iron and aluminum. This profoundly diminishes the short-term fertilizer use efficiency by crops. However, fields that have received P fertilizers over time eventually saturate the highest-energy P-bonding sites in their soils, and consequently a much larger percentage of applied P is maintained in forms available to crops (Syers et al. 2008). If farmers do not decrease fertilization as this saturation point is approached, then soils become over-fertilized, making them highly vulnerable to loss of P from leaching. Syers et al. (2008) refer to the level of readily available soil P at which crop yields approach a maximum and begin to level off as the **critical value.*** They argue that as long as a soil is at or near its critical value, fertilization should approximately equal P-removal rates in harvests, and the leaching hazard of P to water bodies will remain low (see Messiga et al. 2010). This discussion has not addressed P loss from fields via erosion, described in chapter 2, which is globally a greater avenue of P loss from most farms than leaching (Global P erosion is 20 Mtons per year; Smil 2000). If an agricultural field is prone to erosion, then fertilization must equal harvest *plus* erosion losses of P to maintain production in the long run.

Current genetic engineering research that is being conducted to optimize P_i-use efficiency is essentially focused on lowering the "critical P value" of a crop. By enhancing crops' capacities to secrete protons, organic acids, or phosphatases into the rhizosphere, engineers hope to make tightly bound soil P_i reserves more soluble and thus accessible to crops (Shenoy and Kalagudi 2005, Gaxiola et al. 2011). In essence, the pool of readily available soil P is being redefined to include more tightly bonded P.

DEFINING P EFFICIENCY IMPROVEMENTS WITH GE CROPS

Two of the broad dimensions of P sustainability addressed in this volume—the depletion of P fertilizer resources and aquatic eutrophication—would likely be affected by the successful development and planting of crops engineered to more efficiently acquire tightly bound soil-P (**fixed-P***). With regard to easing dependency on finite P fertilizer resources, scenarios for more developed countries and for less developed ones would differ. In countries that have historically applied large amounts of P fertilizers (e.g., the United States and Europe), farmers would enjoy a period of time during which crops would require less or no fertilizer since roots could access previously bound fertilizer P or endogenous P minerals (see chapter 4). But eventually the soil-critical P value would be reached again, as would the same level of crop fertilizer dependency that existed before the introduction of the engineered crop. Thus, the fertilizer P savings would be determined by the magnitude of differences in critical P values between previous non-GE crops and their P-efficient replacements.

In countries or soils that have not historically and will not in the foreseeable future apply P fertilizers, engineered crops with high phosphorus assimilation efficiencies (PAE) would provide a short-term boost in productivity, as crop P limitation is relaxed through appropriation of more tightly bonded P (see chapter 4 for the definition of PAE vs. phosphorus uptake efficiency). However, the enhanced pool of P made accessible by the engineered crops is also finite, and would likely be exhausted in years to decades.

For countries beginning to use P fertilizers on most farmlands, adoption of engineered crops would require less additional P to saturate high-energy bonding sites to reach the critical value. The savings in **non-fixed P*** could be substantial in countries such as Brazil or Indonesia, where intensive production systems are being implemented on highly weathered, P-deficient soils. For example, in a non-GE groundnut cropping system in Indonesia, three P fertilization rates were applied over 25 years.

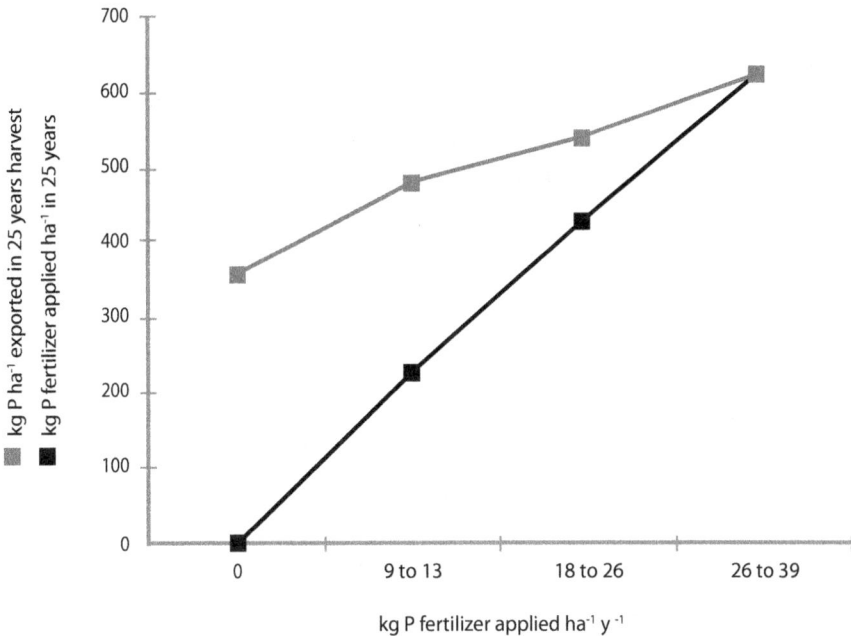

FIGURE 7.3 P exported in groundnut harvests vs. P additions in fertilizer over a 25-year fertilization trial in Indonesia. Units on the x-axis represent range of annual P additions to four fertilization treatments (after Aulakh et al. 2003, Syers et al. 2008).

The highest rate, which amounted to a cumulative application of 621 kg ha^{-1} of fertilizer P, was required to achieve a P balance in which fertilizer P inputs equaled fertilizer P exports in harvests (Figure 7.3; Aulakh et al. 2003; Syers et al. 2008). No transgenic crops featuring enhanced PAE have been developed to date, although current research suggests increases in P uptake efficiencies of >10 percent and potentially much higher are achievable (Yang et al. 2007). If we assume an increase in fertilizer PAE of 10 percent by a genetically engineered groundnut crop, then Indonesian growers might expect to achieve the critical P level in their soils with 62 kg P ha^{-1} less fertilizer than with the non-GE crop. If such a reduction in fertilizer requirements was achieved on all lands dedicated to groundnut production across Indonesia, on the order of 38,602 tons of P could potentially be saved over several years. This would amount to 17 percent of Indonesia's 2010 fertilizer consumption.

REDUCED EUTROPHICATION RISK WITH LOWERED SOIL-CRITICAL P VALUES

In some freshwater ecosystems, P concentrations as low as 0.02–0.035 mg l^{-1} trigger algal blooms and eutrophication (Brookes et al. 1997). The use of concentrated P fertilizers to increase inorganic soluble P for crop production increases the rate of P loss via leaching or erosion from farmland and subsequent risk of eutrophication of freshwater bodies (Sharpley et al. 1997). Transgenic crops with greater PAE would effectively lower the critical P values at which production is optimized and achieve crop yields comparable to those of non-GE varieties at lower concentrations of total and soluble-P. Reducing total and soluble-P_i in agricultural topsoil would directly translate into the reduction of eutrophication hazards associated with many annual cropping systems.

SUSTAINABILITY LIMITATIONS

The values that inform individual and group perspectives on transgenic modification of organisms typically have nothing to do with the element P. Rather, these values tend to be rooted in beliefs pertaining to religion, science, economics, and other human constructs. Nevertheless, these perspectives, and how they translate to policy, may significantly affect the availability of soil-P in agriculture, as well as the eutrophication impacts of P losses from agriculture to aquatic ecosystems. It is important to reiterate, however, that the deployment of transgenic crops with greater PAE than their non-engineered counterparts would only result in a savings of P during an equilibration phase of the soil P to the new crops. In fertilized systems, lowering of the critical P value would require less fertilizer needing to be "fixed" before an equilibrium were reached between fertilizer inputs and crop exports of P. Once the equilibrium is reached, the same fertilizer P inputs would be required to achieve the yields of non-GE crops. For unfertilized crops, farmers would anticipate greater yields for a period of time as the transgenic crops draw down the larger pool of endogenous P they can access. In years to decades, the yields will be expected to revert back to what they were before the introduction of transgenic varieties.

Improvement of PAE in crop plants would reduce leaching of P through soils and subsequent eutrophication of freshwater ecosystems, but losses of P via leaching are minor compared to the more significant problem globally of P losses from soil surfaces via erosion (Smil 2000, Carpenter and Bennett 2011). Crops with higher PAE would not be expected to significantly reduce P losses from erosion.

Genetic engineering is not the only approach that shows promise for developing high PAE crops. Conventional breeding methods can also target improvement in root:shoot ratios or rhizosphere acidification, and moreover, some improvements in PAE have been demonstrated following crop inoculation with plant growth-promoting rhizobacteria (Ramaekers et al. 2010).

In both fertilized and unfertilized scenarios described above, it will be important for farmers to consider that once the decision to grow crops with high PAEs is made, attempts to grow any other crops on the same soils with lower PAEs will probably result in reduced yields. This could effectively lock farmers into only being able to grow a narrow range of crops with high PAE values. This would be true regardless of whether the increase in PAE was achieved through genetic engineering or conventional breeding.

The controversy over genetic engineering is a fascinating example of how ethical values influence societal choices that ultimately carry biogeochemical consequences. Simply put, the global fluxes of P will be affected, likely significantly, by which values predominate in international policy stances toward genetically engineered crops and livestock. Consequently, global P fluxes should inform the policy debate on GE crop deployment.

DIET CHOICES AND P BIOGEOCHEMISTRY

Another important way in which cultural norms and values affect the global P cycle is through food choices. The food we eat not only contains varying amounts of P and other elements, but food choices also strongly affect P incorporation efficiency, with vegetarian/plant-based diets differing substantially from omnivorous diets that include meat. With ecological theory as a guide, we know that adding a trophic level in a food web decreases the efficiency at which carbon and nutrients are delivered to the highest trophic level from the primary producers, often by a factor of 10 for each trophic step. We address this issue from the P-perspective, but food choice also has important effects on the biogeochemistry of many elements and molecules from P, nitrogen, and carbon to water.

THE IMPLICATIONS OF OMNIVORY VS. VEGETARIANISM TO P BIOGEOCHEMISTRY

In the past 10,000 years, the development of agriculture and changes in land use have significantly modified human diets. Early in human evolution, human ancestors in Africa are thought to have converted from an overwhelmingly plant-based diet to one dominated by meat as climate change shifted the vegetation from dense edible trees to open savannah (Leonard 2002). Indeed, feedbacks among a meat-intensive diet, cooked food, substantial increases in brain size, and the advent of complex social organizations (including hunting parties) may have been central to defining the intellectual and social capacities of our species (Leonard 2002).

While intensive meat consumption may have facilitated the expansion of hominid brains, early *Homo sapiens* effectively employed their increased intellectual capacity to develop a wide range of diets based on food resources available and dietary needs in newly colonized ecosystems. Thus, diets of indigenous peoples around the Earth have ranged from overwhelmingly vegetarian to overwhelmingly carnivorous. With the advent of agriculture, diets tended to shift away from meats, vegetables, fruits, and roots, to high-caloric grains, dairy products, refined sugars, and oils (Eaton and Konner 1985). Agriculture also led to a decrease in the diversity of the foods that humans eat. Not only has the choice of foods changed, the nutritional composition of individual foods has as well. For instance, meat from wild bison, deer, and horses was much less fatty than the meat we buy at a supermarket today.

Today, about 30 percent of the world's population lives on a meat-based diet and the remaining two-thirds of the population eat primarily a plant-based diet (Pimentel and Pimentel 2003). There are more than 65 billion domesticated animals on Earth today, which, if equally divided, translates into nearly 10 for every human (Childers et al. 2011). History has shown that in most cultures, when affluence increases, so does the consumption of meat. Recent estimates hold that global beef production will increase by about 1500 metric tons and chicken by 5000 metric tons in the next 20 years alone (Fiala 2008).

RELIGIOUS NORMS AND MEAT CONSUMPTION

Diet is a major determinant of P-use in agriculture, and cultural values and religious beliefs are major determinants of diet. Religious beliefs can motivate people to consume greater quantities of meat (Johnson et al. 2011), or, conversely, influence people to eat less. For example, vegetarianism is part of the practice of several Indian religions. Jainism holds that all sentient beings are sacred and therefore vegetarianism is mandatory. While a plant-based diet is not obligatory throughout Hinduism, important scriptures dictate various forms of it. The *Mahabharata* (3.199.11–12; 13.115; 13.116.26; 13.148.17) and the *Laws of Manu* (5.27–44) condemn animal slaughter and specify vegetarian practice. Violence to animals entails negative karmic influences, and eating flesh, or cutting, buying, selling, cooking, or serving it, binds man to suffering (Simoons 1994). Many Hindus follow the prescriptions of the *Bhagavad Gita* (3.13) and offer food as *prasad*, for which only vegetarian dishes are appropriate. The vegetarian diet associated with the *sattvic* qualities essential for spiritual discipline is important in the Hatha Yoga tradition.

According to a recent survey (Yadav and Kumar 2006), the contemporary Hindu diet varies according to community or caste and regional traditions. While the majority of Hindu vegetarians are lacto-vegetarians who do not eat eggs, those who live on the coast are far more likely to consume fish than those in the interior. More than half of Hindu Brahmins (Yadav and Kumar 2006), and nearly a third of the overall Indian population do not eat meat at all. In 2009 the Indian Environment Minister commented on global beef consumption patterns in India, China, and the

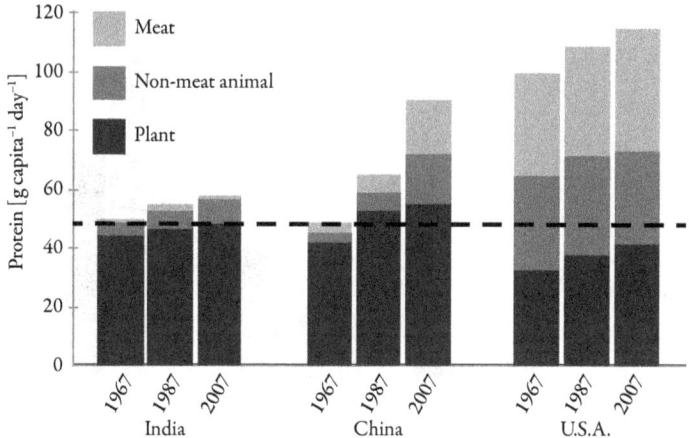

FIGURE 7.4a Changes in daily per capita protein consumption from 1967 to 2007 in India, China, and the United States. The dashed line represents the FAO daily consumption recommendation. (*Source:* FAOSTAT 2011)

United States and the methane emissions associated them, saying that "the solution was to stop eating beef. The best thing for us, India, is that we are not a beef eating nation" (see Nelson 2009). Overall, while India has a wide diversity of traditions with varying degrees of dietary restrictions, it is clear that the Hindu religious values have limited consumption of meat in general and beef in particular (Figure 7.4a).

In 65 C.E., Buddhist monks migrated from India, where their teachings on plant-based diets had complemented those of Hinduism, to China, where other views on diet came to the fore. Despite the prohibition on killing, the Buddha had refused to dictate dietary food rules. When a disciple named Mahamati asked him to "teach us as to the merit and vice of meat-eating," the Buddha replied at length. In the Mahayana Mahaparinirvana and Lankavatara Sutras, the Buddha calls for a compassionate vegan lifestyle but predicts that later monks, as they beg for food during their migrations, would claim that the Buddha allows the eating of meat (Phelps 2004). The diverse ethnic groups that comprise modern-day China have fewer restrictions on the consumption of beef, or meat in general, than the peoples of India, and in recent decades China's consumption of meat has grown in concert with its economy (Figure 7.4a–7.4c).

The Jewish religion has many dietary traditions, including restrictions on the consumption of particular animals, organs, animal products, or methods of preparation (Rich 2011). The Torah (Lev. 11:3; Deut. 14:6) forbids consumption of mammals that are not ruminants or cloven-hoofed (such as camel, hare, and pig), whereas cattle, sheep, goats, deer, and bison are acceptable (Rich 2011). Islam forbids the consumption of pork and blood (Qu'ran 2: 173) and prescribes specific practices for animal slaughter and meat preparation (Qu'ran 5:3).

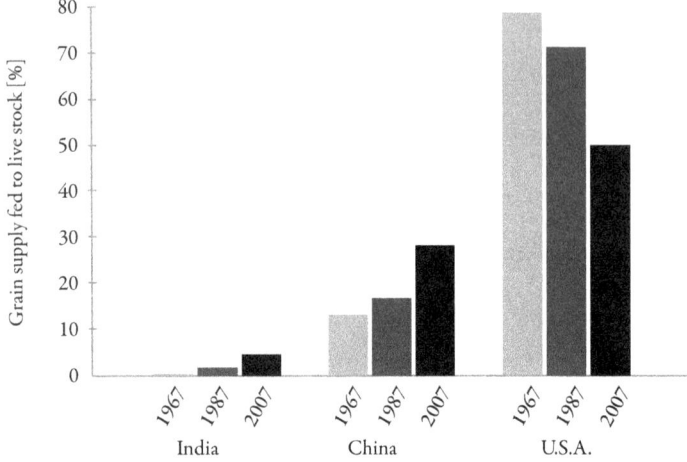

FIGURE 7.4b Changes in percent of grain supply fed to livestock in India, China, and the United States from 1962 to 2002.

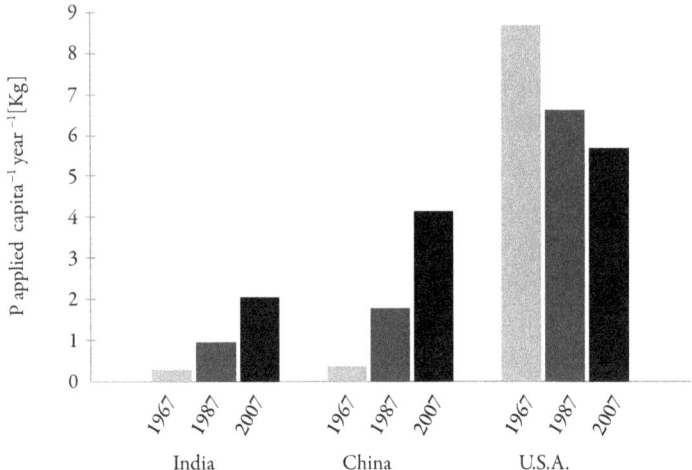

FIGURE 7.4c Changes in P fertilizer applied to cropland per capita in India, China, and the United States from 1967 to 2007 (*Source:* FAOSTAT 2011).

Christianity has among the fewest dietary restrictions of any major religion, a freedom of diet that resulted in part from early Christendom's rejection of Jewish dietary restrictions (Coveney 2000). In the New Testament, Peter is quoted as saying "What God has made clean, you have no right to call profane" (Acts 10:15). That Christian diets are generally unencumbered by religious laws or customs has allowed per capita meat and animal product consumption levels to soar in the United States and other predominantly Christian countries (Figure 7.4a).

While meat consumption is governed by a wide range of economic and agricultural resource factors, religious-based values have influenced the strikingly different

P budgets for the Indian, Chinese, and U.S. populations. Meat production in the United States is based on grain as feed, and China has increasingly adopted this production model as its economy has grown in recent decades (Figure 7.4b). Increased demands for grain directly translate into increased applications of P to support their production (Figure 7.4c).

The amount of P fertilizer used annually in the United States and India is not very different: about 4200 and 5500 metric tons, respectively (Heffer 2009). If we assume a 10 percent trophic transfer efficiency, it means that only 5.5 percent of the P in fertilizer applied in the United States is actually assimilated by humans, while in India about 9.6 percent of the P applied as fertilizer is assimilated. So Indians, just by eating lower on the food chain, increase their P-use efficiency by 75 percent relative to people in the United States. Others have suggested that the difference between the amount of P required for vegetarian vs. an omnivorous diet is even greater, with meat-based diets requiring about three times as much P as a vegetarian diet (Schröder et al. 2011). It is worth noting that globally we mine about 14 Mt per year of P from the Earth for fertilizers and if we were to provide all 7 billion people on the planet with the USDA-recommended amount of P per day (ca. 640 mg), it would require about 1.6 Mt of P. Therefore, an efficiency of P transfer into humans closer to that of India, rather than that of the United States, could actually provide all of humanity with its P needs. But the current trend is that increasing numbers of the world's people are consuming more, not less, meat in their diet.

THE USE OF HUMAN EXCRETA IN AGRICULTURE

Culture and P biogeochemistry also intersect in the management of human excreta. A full-grown adult excretes a similar amount of P as is contained in the food he or she ingests. Cordell et al. (2009) report that vegetarians excrete approximately 0.3 kg P capita^{-1} yr^{-1} in urine plus feces, while people with meat-intensive diets excrete 0.6 kg P capita^{-1} yr^{-1}. If we assume a global average of 0.4 kg P capita^{-1} yr^{-1}, the P in human excreta amounts to 2.78 Mt P yr^{-1}, which is 20 percent of the total P produced globally in fertilizers each year (Cordell et al. 2009).

While most agricultural societies fertilize some croplands with animal manure, the practice of applying night soil, or "humanure," on crops has become less common, even with great advances in pre-application treatments to establish pathogenic safety (Drangert 1998, George 2008). The relative ease of handling and transport generally makes synthetic fertilizers preferable to human and animal manures, but **cultural stigmas*** surrounding the use of human excreta also affect the extent to which it is used as fertilizer (*see* **fecophobe*** and **fecophile***).

HISTORICAL, CULTURAL, AND RELIGIOUS FACTORS IN EXCRETA HANDLING AND REUSE

The Chinese have applied excreta to farm fields for four thousand years (King 1911) and continue to use significant quantities as a soil amendment (see **fecophile***). Pan et al. (1995 as cited in Liu et al. 2008) report that 94 percent of excreta from rural areas—although under 30 percent from urban areas—was applied to farmlands in the 1990s (Chen 2002 as cited in Liu et al. 2008). If one assumes an annual per capita excreta-P mass of 0.4 kg yr^{-1} for the entire Chinese population (based on Cordell 2009), then the amount of P recycled into agriculture through excreta in the 1990s was approximately 0.34 Mt yr^{-1}. While far less than the 8.91 Mt of P applied in China as fertilizer in 1995, the excreta-P would nevertheless be sufficient to fertilize 10 percent of permanent, arable Chinese cropland with 22 kg ha^{-1} yr^{-1} (FAO 1995, 2011)—an adequate P application rate for a high-yielding wheat crop (Newman 1997). Similarly, Japan inherited the reuse tradition from China a millennium ago (King 1911) and has municipalities that still recycle waste to farm use (Drangert 1998).

In Africa, by contrast, attitudes toward human excreta are quite negative (see **fecophobe***). Feces may be used as a warning to invoke fear in others (Winblad and Kilama 1985) or considered lethal and used in witchcraft. In some regions, simple verbal reference to excreta is taboo (Van Der Geest 1998). Similar reluctance to handle human waste is typical of large parts of South Asia that have been influenced by Hinduism and Islam. Speigel (2011) noted that "psychological contagion" threatens the economic viability of crops to which excreta-derived fertilizers have been applied.

Western cultures gradually have become isolated from excreta with urban development and the introduction of sanitary sewers. In the late eighteenth century, London and Paris built massive installations designed to carry stormwater but not household wastes (George 2008; Luthi et al. 2011). Latrines and cesspools containing urine and feces were managed separately and their contents applied on nearby agricultural lands (Barles 2007). Highly decentralized networks, what Joel A. Tarr calls the "cesspool-privy vault-scavenger system" (cited in Melosi 2000), functioned adequately to transfer fertilizer to fields until population pressures led to leaching and accidental cesspool overflows (Barles 2007). In urban Europe and America, sewers were further overburdened with the introduction of water into the home. Following John Snow's demonstration of germ theory (Johnson 2007), sanitary pipes were laid, often adjacent to original storm sewers, and eventually carried the contents of flush toilets, feeding into the original infrastructure before flowing into nearby bodies of water (Melosi 2000).

By the end of the nineteenth century it was becoming less and less socially acceptable to share a privy or carry one's chamber pot to be emptied. As social mobility ramped up in the Victorian period, class differences appeared in the degree of exclusivity of the toilet room and the quality of the hardware. Embraced by the then-more-egalitarian society in the New World, the flush toilet carried this culture of fecal aversion straight into American homes. The flush-and-forget reality of Western and urban sanitation practice means that people rarely if ever had anything to do with managing the disposal of their own feces, reinforcing their disgust and perpetuating what Praeger (2007) calls "fecal denial."

Religious prescriptions regarding excretion affect social practice across time and space. For Hindus, elaborate rules for defecation in the 1500-year-old scripture *Manusmriti Vishnupuran* specify special mantras and dress followed by washing of the anal area using the left hand. The *Vishnu Purana* outlines post-defecation practice to cleanse one's body and clothing and gives environmentally sound guidance on where elimination is permissible (Pathak 2011). Hindu requirements regarding proximity to excreta vary according to caste, with night soil collection traditionally the work of Dalits, or "unscheduled" castes (George 2008). The Qur'an (5:6) specifies a purification protocol following defecation and various other activities considered unclean. The Islamic Hadith report the hygiene practices of the Prophet and specify that while defecating outdoors Muslims face neither directly toward nor away from Mecca. While the Bible counsels defecating away from the camp and covering feces (Deuteronomy 23:10–14), Jews and Christians have scant scriptural guidance on elimination, and nothing whatsoever on excreta reuse. At the other end of the spectrum, Buddhism embraces the cycles of nature, including the recycling of human excreta, as a spiritual concept (Avvannavar and Mani 2008).

Avvannavar and Mani (2008) have classified peoples on the basis of their association with excreta. Fecophiles see human excreta as part of a natural cycle and use it in agriculture. Like the Chinese and Japanese, these societies have traditionally lived in dense settlements, cooked their food, and avoided eating raw vegetables. In contrast, fecophobic peoples avoid physical contact with human excreta and are uncomfortable talking about it. Among them are Arabs and some Africans, who are heirs to semi-nomadic lifestyles with little familiarity with latrine construction or the cultivation of crops that could be fertilized (Winblad and Kilama 1985). Childers et al. (2011) note that "overcoming this psychological stigma is necessary and far from trivial" if recycling is to be explored.

GLOBAL IMPERATIVES AND SUSTAINABLE SANITATION PRACTICES

The development of water-based sanitation infrastructure was a milestone, declared "the greatest medical advance" in 166 years by the *British Medical Journal* (2007). The mixing of industrial and household wastes in sewers, however, has introduced new health risks (Dorfman et al. 2004, Corcoran et al. 2010) and also has resulted in an estimated annual net loss of 10.5 million Mt of phosphorus from the world's croplands (Liu et al. 2008). Deteriorating sewage infrastructure, energy costs, climate change pressures, and the need to recover P for food security (Cordell et al. 2011) are provoking a rethinking of sanitation that goes directly to alternatives that close sanitation loops (Sindani 2011).

With half the current global population living in cities and increasingly scarce fossil fuels and fertilizers threatening food security, there is growing popular appreciation of the benefits of locally available soil amendments. At the same time, the limits and the financial and environmental costs of water-based sanitation are better understood (Langergraber and Muellegger 2005, Grunbaum 2010). In Western countries a new acceptance of excreta reuse is seen among those involved in permaculture design (Holmgren 2003), urban agriculture (Allen and Conant 2010), and green building (Cascadia Green Building Council 2011a and b).

As nations struggle toward the Millennium Development Goals (MDGs) objective to "halve, by 2015, the proportion of the population without sustainable access to safe drinking water and basic sanitation," 2.6 billion people remain without "improved sanitation" (WHO/Unicef/JMP. 2008). Improved sanitation includes "flush-and-discharge" technologies such as sewered or septic systems, "drop-and-store" methods (Drangert 1998, Winblad and Simpson-Hebèrt 2004) such as "cat hole" defecation, pit latrines, and Ventilated Improved Pit (VIP) latrines. Only one of these options, composting toilets, has significant potential for the recycling of phosphorus back into food production (Rosemarin et al. 2008, Schröder 2011).

Over the past decade, scholars and practitioners have articulated alternatives to conventional "end-of pipe" technologies (Luthi et al. 2011, Bracken et al. 2007). One approach that holds considerable promise for the cycling of P involves the separation of urine from feces. Diverting urine in a dry toilet system can prevent bacteria in feces from releasing ammonia, which causes odor, and makes possible its use directly on crops, usually diluted with water. Diverting urine in a flush toilet and piping it to a decentralized location facilitates use in agriculture either directly or through the production of struvite ($NH_4MgPO_4 \cdot 6H_2O$), and it allows for greater control of water pollution (Larsen et al. 2009). Although humans typically produce urine in 10 times the volume of feces, the urine is relatively easy to recycle for its high N and P value. It makes more sense, however, to compost and sanitize pathogenic feces in containers or fenced areas near residences. Separated feces dry quickly, shrink in volume, and can be easily and safely handled in a year or less. Evidence from controlled research and field testing shows that dry treatment of the feces following urine separation destroys pathogens more effectively than mixed composting methods (Dellstrom Rosenquist 2005).

While separation at the source optimizes opportunities to capture and recycle P from both urine and feces, adoption may present "problems regarding the social acceptability in faecophobic societies or negative effects due to misuse" of unfamiliar toilet technologies (Schröder 2011). After all, "the source" here is human beings in the act of elimination, an action so private that it is barely discussed in most societies. Reinventing sanitation, moreover, confronts entrenched ways of thinking and acting that extend far from the toilet. Ecologically sound sanitation is less a matter of technical solutions and more what Drangert (1998) calls "an intriguing interplay of norms and attitudes among professionals as well as users."

The challenges of excreta logistics are considerable worldwide. Schröder (2011) has mapped out the logistics chain for an African city and highlighted cultural attitudes that come into play at every step. Negative attitudes of slum residents supplying excreta, employees of the private company handling logistics, and farmers purchasing the sanitized product were documented and mitigation through incentives and sensitization recommended. To overcome these constraints, he also calculated economic benefits that resonate with policy makers. Benefit:cost ratios for Kampala, Uganda, show that investments in sustainable sanitation of 4.37 EUR would avert the economic burden of 28.05 EUR in repair of existing infrastructure on an annual basis. He further pointed out that logistically appropriate recycling of excreta advances nutrition both by augmenting harvests and by protecting people from diarrhea, via contaminated water sources, that inhibits effective uptake of the nutrients in food (Schroder 2011).

Cultural attitudes about the handling and management of human excreta are deeply ingrained. Behaviors that avoided the reuse of excreta in agriculture and served people well in certain circumstances in the past may no longer be rational given advances in technology and the science of pathology (Curtis et al. 2004). Nevertheless, fecophobia presents one of the greatest challenges to closing the P cycle in agriculture. Of considerable interest are a growing number of examples where cultural attitudes toward excreta recycling have shifted. Adoption of reuse-oriented ecological sanitation technologies, both in the world's more developed (Etter et al. 2011, Braum 2011, Mariwah and Drangert 2011, Robinson 2005, Schröder 2011) and less developed regions (Fall and Coulibaly 2011, Larsen et al. 2009, Winker 2010) demonstrate that cultural constraints to excreta recycling can be overcome.

SOCIAL VALUES AND BIOGEOCHEMICAL CYCLES

In this chapter, we have provided three illustrations of how cultural beliefs, values, and traditions can affect P sustainability. We chose examples in which cultural norms drive demand for P (diet), crop assimilation efficiency of P (genetic engineering), and return of P to the soil (excreta use), but there are numerous other examples in which cultural choices or behaviors translate into greater or lesser P demands. For example, a society's choice or ability to regulate human population size, or to consume alcohol (either as beverage or fuel), which requires considerable land and inputs to produce, or willingness to implement municipal composting, all carry P sustainability ramifications.

To a significant extent, the challenge of achieving a more sustainable society in general, and a more sustainable relationship with P in particular, involves the reconciliation of social behaviors that conflict with scientific insights. Such reconciliation can be particularly challenging when the linkages between cultural norms and sustainability issues are not even perceived by the relevant social groups (i.e., P depletion and meat consumption in the United States). It is with this challenge in mind that we underscore the importance of broad-based, democratic participation in defining and acting on sustainability goals (chapter 9). Developing complex and multi-scaled solutions, such as those required to address the human P dilemma, will necessitate inclusion and engagement of people whose behaviors and traditions help drive current P-use patterns.

BOX 7.2
CHAPTER 7 SUMMARY

- We examined three human-culture interactions with P biogeochemistry: a) genetically engineered organisms, b) human diet, and c) management of human excreta.
- Genetically engineered plants may be important tools for increasing the P available to crops and decreasing the quantity of P fertilizer applied by farmers, particularly in the short term. The societal choice of whether to adopt or reject GE plants will impact global reserves of affordable P.
- Food preferences rank among the most deeply ingrained of cultural attributes. While the element phosphorus was not considered in the evolution of cultural norms regarding the consumption of meat, cultures with meat-intensive diets require far greater P resources to maintain their existence.
- A wide range of religious, aesthetic, and historical reasons influence whether a society is more "fecophobic" or "fecophilic." Cultures that accept the use of excreta in agriculture effectively close a critical part of the P nutrient cycle and reduce dependence on finite P resources.
- It is important to recognize that deep-seated human values and cultural traditions can inadvertently drive the demand for and fate of phosphorus. The question of how to evolve cultural norms is as sensitive as it is essential in devising approaches to improve P-use sustainability.

Acknowledgments

We are grateful to the organizers of the P Sustainability Summit, including J. Corman, K. Wyant, and J. Elser. Without their dedication and efforts, we could not have written this paper. J. Cotner acknowledges the support of NSF (DEB 0918753). S. Moratto provided research assistance.

REFERENCES

Allen, L., and J. Conant. 2010. Backyard urine recycling in the United States of America: An assessment of methods and motivations. *Sustainable Sanitation Practice* 2010: 3http://www.ecosan.at/ssp/issue-03-use-of-urine/issue-03

Aulakh, M. S., B. S. Kabba, H. S. Baddesha, G. S. Bahl, and M. P. S. Gill. 2003. Crop yields and phosphorus fertilizer transformations after 25 years of applications to a subtropical soil under groundnut-based cropping systems. *Field Crops Research* 83: 283–296.

Avvannavar, S. M., and M. Mani. 2008. A conceptual model of people's approach to sanitation. *Science of the Total Environment* 390:1–12.

Barles, S. 2007. Urban metabolism and river systems: an historical perspective—Paris and the Seine, 1790–1970. *Hydrology and Earth System Sciences Discussions* 4: 1845–1878.

Bracken, P., A. Wachtler, A. R. Panesar, and J. Lange. 2007. The road not taken: how traditional excreta and greywater management may point the way to a sustainable future. *Water Science & Technology: Water Supply* 7:219–227.

Braum, C. 2011. Economic studies: Economic efficiency—MAP vs. Urine. Report to meeting of SANIRESCH, February 2011. http://www.saniresch.de/images/stories/downloads/Summary-02-2011-en.pdf

British Medical Journal 2007. Medical Milestones: Celebrating Key Advances Since 1830. http://www.bmj.com/content/suppl/2007/01/18/334.suppl_1.DC2/milestones.pdf

Brookes, P. C., G. Heckrath, J. Smet, G. Hofman, J. Vanderdeelen, H. Tunney, O. T. Carton, and A. E. Johnston. 1997. Losses of phosphorus in drainage water. In: *Phosphorus Loss from Soil to Water*, edited by H. Tunney, O. T. Carton, P. C. Brookes, and A. E. Johnston, 253–272. CAB International, Wallingford, UK.

Carpenter, S. R., and E.M. Bennett. 2011. Reconsideration of the planetary boundary for phosphorus. *Environmental Research Letters* 6:014009.

Cascadia Green Building Council. 2011a. Regulatory Pathways to Net Zero Water. http://cascadiagbc.org/resources/RegulatoryPathwaystoNetZeroWater.pdf

Cascadia Green Building Council. 2011b. Toward Net Zero Water: Best Management Practices for Decentralized Sourcing and Treatment. http://cascadiagbc.org/resources/TowardNetZeroWater.pdf

Chen, F. 2002. *Agricultural Ecology*. China Agricultural University Press. (In Chinese.)

Childers, D., J. Corman, M. Edwards, and J. J. Elser. 2011 Sustainability Challenges of Phosphorus and Food: Solutions for Closing the Human Phosphorus Cycle. *BioScience* 61:117–124.

Corcoran, E., C. Nellemann, E. Baker, R. Bos, D. Osborn, H. Savelli (eds). 2010. Sick Water? The central role of waste-water management in sustainable development. A Rapid Response Assessment. United Nations Environment Programme, UN-HABITAT, GRID- Arendal. http://www.grida.no

Cordell, D., J. O. Drangert, and S. White. 2009. The story of phosphorus: Global food security and food for thought. *Global Environmental Change* 19:292–305.

Cordell, D., A. Rosemarin, J. J. Schröder, and A. L. Smit. 2011. Towards global phosphorus security: A systems framework for phosphorus recovery and reuse options. *Chemosphere* 84:747–758.

Coveney, J., 2000. *Food, Morals and Meaning: The Pleasure and Anxiety of Eating*. Routledge, Florence, KY.

Crews, T. E., K. Kitayama, J. H. Fownes, R. H. Riley, D. A. Herbert, D. Mueller-Dombois, and P. M. Vitousek. 1995. Changes in soil phosphorus fractions and ecosystem dynamics across a long chronosequence in Hawaii. *Ecology* 76:1407–1424.

Curtis, V., Aunger, R. & Rabie, T. 2004. Evidence that disgust evolved to protect from risk of disease. *Proceedings of the Royal Society of London*, Series B, 271 (Supplement), 131e133.

Dellstrom Rosenquist, L. E. 2005. A psychosocial analysis of the human-sanitation nexus. *Journal of Environmental Psychology* 25:335–346.

Dorfman, M., N. Stoner, and M. Merkel. 2004. Swimming in sewage. Natural Resources Defense Council and the Environmental Integrity Project, New York, NY [Online.] http://www.nrdc.org/water/pollution/sewage/sewage.pdf

Drangert, J-O. 1998. Fighting the urine blindness to provide more sanitation options. Water SA 24:2, 157.

Drees, W. B., ed. 2011. The Heythrop Journal. Wiley Online Library. http://onlinelibrary.wiley.com/doi/10.1111/j.1468-2265.2011.00646_55.x/full

Eaton, S. B., and M. Konner. 1985. Paleolithic nutrition. *New England Journal of Medicine* 312:283–289.

Edwards, M. G., G. M. Poppy, N. Ferry, and A. M. R. Gatehouse. 2009. Environmental benefits of genetically modified crops. In *Environmental Impact of Genetically Modified Crops*, edited by N. Ferry and A. M. R. Gatehouse. 23–41 CAB International.

Etter, B., E. Tilley, and K. M. Udert. 2011 Urine as a Resource in Nepal. Sandec News, July 2011. 12:16.

Fall, A., and C. Coulibaly (Draft). 2011. *Urban urine diversion dehydration toilets and reuse Ouagadougou*, Burkina Faso. http://www.susana.org/lang-en/case-studies?view=ccbktypeitem&type=2&id=84

FAO 1995, 2007, 2011. Food Agriculture Organization. http://faostat.fao.org/site/339/default.aspx. FAO-STAT.

Fiala, N. 2008. Meeting the demand: An estimation of potential future greenhouse gas emissions from meat production. *Ecological Economics* 67:412–419.

Gaxiola, R., M. Edwards, and J. J. Elser. 2011. A transgenic approach to enhance phosphorus use efficiency in crops as part of a comprehensive strategy for sustainable agriculture. *Chemosphere* 84(6):840–845.

George, R. 2008. *The Big Necessity: Adventures in the World of Human Waste*. Portobella Books, London.

Grunbaum. 2010. Gee whiz: Human urine is shown to be an effective agricultural fertilizer. *Scientific American* 303:1.

Heffer, P. 2009. Assessment of Fertilizer Use by Crop at the Global Level 2006.

Holmgren, D. 2003. *Permaculture: Principles and Pathways Beyond Sustainability*. Holmgren Design Services, Victoria, Australia.

ISAAA. 2010. http://www.isaaa.org/resources/publications/briefs/42/executivesummary/default.asp

Johnson, K. A., A. E. White, B. M. Boyd, and A. B. Cohen. 2011. Matzah, Meat, Milk, and Mana. *Journal of Cross-Cultural Psychology* 42:1–16.

Johnson, S. 2007. *The Ghost Map*. Riverhead Books, New York.

King, F. H. 1911. *Farmers of Forty Centuries*. Rodale Press, Emmaus, PA.

Kumwenda. 2011. Analysis: More African countries seen growing GM crops. www.reuters.com.

Langergraber, G., and E. Muellegger. 2005. Ecological sanitation—a way to solve global sanitation problems? *Environment International* 31:433–444.

Larsen, T. A., A. C. Alder, R. I. L. Eggen, M. Maurer, and J. Lienert. 2009. Source separation: Will we see a paradigm shift in wastewater handling? *Environmental Science & Technology* 43:6121–6125.

Leonard, W. R. 2002. Food for thought. *Scientific American*. December 287:106–115.

Liu, Y., G. Villalba, R. U. Ayred, and H. Schroder. 2008. Global phosphorus flows and environmental impacts from a consumption perspective. *Journal of Industrial Ecology* 12:229–247.

Luthi, C., A. Panesar, T. Schütze, A. Norström, J. McConville, J. Parkinson, D. Saywell, and R. Ingle. 2011. Sustainable sanitation in cities—A framework for action. Sustainable Sanitation Alliance and International Forum on Urbanism. Papiroz Publishing House, Rijswijk, The Netherlands. http://www.eawag.ch/forschung/sandec/publikationen/sesp/dl/sustainable_san.pdf

Mariwah, S., and J-O Drangert. 2011. Community perceptions of human excreta as fertilizer in peri-urban agriculture in Ghana. *Waste Management and Research* 29(8): 815–822.

Melosi, M. V. 2000. *The Sanitary City: Urban Infrastructure in America from Colonial Times to the Present*. Johns Hopkins University Press, Baltimore, MD.

Messiga, A. J., N. Ziadi, D. Plenet, L. E. Parent, and C. Morel. 2010. Long-term changes in soil phosphorus status related to P budgets under maize monoculture and mineral P fertilization. *Soil Use and Management* 26:354–364.

Nelson, D. 2009. *India tells West to stop eating beef*. Nov 20, 2009 *The Telegraph* (London).

Newman, E.I. 1997. Phosphorus balance of contrasting farming systems, past and present. Can food production be sustainable? *Journal of Applied Ecology* 34:1334–1347.

Nicholl, D. S. T. 2008. *An Introduction to Genetic Engineering*, 3rd edition. Cambridge University Press, Cambridge, UK.

Paarlberg, R. L. 2009. *Starved for Science*. Harvard University Press, Cambridge, MA.

Pan, S. C., G. H. Xu, Y. Z. Wu, and J. H. Li. 1995. A background survey and future strategies of latrines and nightsoil treatment in rural China. *Journal of Hygiene Research* 24 (Sup): 1–10. (In Chinese.)

Pathak, B. 2011. Sulabh International Museum of Toilets. http://www.sulabhtoiletmuseum.org/fact.htm

Phelps, N. 2004. *The Great Compassion: Buddhism & Animal Rights*. Lantern Books, New York.

Pimentel, D., and M. Pimentel. 2003. Sustainability of meat-based and plant-based diets and the environment. *American Journal of Clinical Nutrition* 78: 660S–663S.

Praeger. D. 2007. *Poop culture: How America Is Shaped by Its Grossest National Project*. Feral House, Los Angeles.

Ramaekers, L., R. Remans, I. M. Rao, M. W. Blair, and J. Vanderleyden. 2010. Strategies for improving phosphorus acquisition efficiency of crop plants. *Field Crops Research* 117:169–176.

Rich, T. R. 2011. Judaism 101: Kashrut: Dietary Laws. http://www.jewfaq.org/kashrut.htm

Robinson, B. E. 2005. Household Adoption of Ecological Sanitation. Masters Thesis. Massachusetts Institute of Technology. http://web.mit.edu/watsan/Docs/Student percent20-Theses/Kenya/Robinson_Ecosan_Thesis_2005.pdf

Rosemarin, A. N. Ekane, I. Caldwell, E. Kvarnström, J. McConville, C. Ruben, M. Fodge. 2008. Pathways for Sustainable Sanitation—Achieving the Millenium Development Goals. SEI/IWA, Stockholm/London.

Schröder, E. 2011 Economic Effects of Sustainable Sanitation Logistics of Human Excreta in Uganda. Diploma Thesis, Christian-Albrechts-University, Kiel, Germany, accessed 09.30.11 http://www.susana.org/docs_ccbk/susana_download/2-1221-schroeder-economic-effects-of-sustainable-sanitation.pdf

Schröder, J. J., A. L. Smit, D. Cordell, and A. Rosemarin. 2011. Improved phosphorus use efficiency in agriculture: A key requirement for its sustainable use. *Chemosphere* 84:822–831.

Sharpley, A. N., S. Rekolainen, H. Tunney, O. T. Carton, P. C. Brookes, and A. E. Johnston. 1997. Phosphorus in agriculture and its environmental implications. In *Phosphorus Loss from Soil to Water*, edited by H. Tunney, O. T. Carton, P. C. Brookes, and A.E. Johnston, 1–53. CAB International, Wallingford, UK.

Shenoy, V. V., and G. M. Kalagudi. 2005. Enhancing plant phosphorus use efficiency for sustainable cropping. *Biotechnology Advances* 23:501–513.

Shiva, V. 2000. *Stolen Harvest*. South End Press, Cambridge, MA.

Simoons, J. S. 1994. *Eat Not This Flesh: Food Avoidances from Prehistory to the Present*. University of Wisconsin Press, Madison, WI.

Sindani, E. Y. 2011. *Linking agriculture and sanitation: How business-driven partnerships can create low-cost services to the urban poor. Sustainable Sanitation Design*. Technical Café AfricaSan 3, July 19–22, 2011, Kigali, Rwanda.

Smil, V. 2000. Phosphorus in the environment: Natural flows and human interferences. *Annual Review of Energy and the Environment* 25:53–88.

Speigel, A. 2011. Why Cleaned Wastewater Stays Dirty In Our Minds. NPR. http://www.npr.org/2011/08/16/139642271/why-cleaned-wastewater-stays-dirty-in-our-minds

Stewart, C. N., M. D. Halfhill, and S. I. Warwick. 2003. Transgene introgression from genetically modified crops to their wild relatives. *Nature Reviews Genetics* 4: 806–817

Sydney Morning Herald. 2011. Bolivia switches on modified food ban. *Sydney Morning Herald*, Australia, June 7, 2011.

Syers, J. K, A. E. Johnson, and D. Curtin. 2008. *Efficiency of soil and fertilizer phosphorus use*. FAO Fertilizer and Plant Nutrition Bulletin no. 18. Rome.

Van Der Geest, S. 1998. Akan Shit: Getting Rid of Dirt in Ghana. *Anthropology Today* 14:3, 8–12.

WHO/UNICEF/JMP. 2008. World Health Organization/United Nations Children's Fund. Joint Monitoring Programme for Water Supply and Sanitation). http://www.wssinfo.org/about-the-jmp/introduction/

Winblad, U., and W. Kilama. 1985. *Sanitation without Water*. Macmillan, London.

Winblad, U., and M. Simpson-Hèbert (Eds.). (2004). Ecological Sanitation. Stockholm: Stockholm Environment Institute. http://www.ecosanres.org/pdf_files/Ecological_Sanitation_2004.pdf; accessed 09.30.11

Winker, M. 2010. Are pharmaceutical residues in the urine a constraint on the use of urine as fertilizer? *Sustainable Sanitation Practice* 3:18–24.

Yadav, Y., and S. Kumar. 2006. The food habits of a nation. *The Hindu*. August 14, 2006.

Yang, H., J. Knapp, P. Koirala, D. Rajagopal, W. A. Peer, L. K. Silbart, A. Murphy, and R. A. Gaxiola. 2007. Enhanced phosphorus nutrition in monocots and dicots over-expressing a phosphorus-responsive type I H+-pyrophosphatase. *Journal of Plant Biotechnology* 5:735–745.

8

How MFA, Transdisciplinarity, Complex Adaptive Systems Thinking, and Education Reform Are Keys to Better Managing P

P IS FOR PARITY

Rebecca Cors, Kazuyo Matsubae, Anita Street

Angela Cazel Jahn, Frances McMahon Ward and Cory Dunnington
Our Floating Days

Scientific Collaborators:
James Elser, Professor, School of Life Sciences, Arizona State University
Roberto Gaxiola, Professor, School of Life Sciences, Arizona State University
Marcia Kyle, Senior Research Associate, School of Life Sciences, Arizona State University

Description of Artwork:
Our Floating Days is a whimsical metaphor of how we entertain uncertainty and vague premonitions of disruption or change. While big problems like peak oil, climate change, and phosphorus scarcity loom ahead, most of us live our everyday lives immersed in the routine and circumstance of our time and place: It can be beautiful here. Unprecedented numbers of people have water, air, food, shelter, beauty, culture. A while later, the same environment made of the same materials might look and function very differently.

Early phosphorus use in agriculture was the happy consequence of erosion. As humans learned to raise crops more efficiently, we began to supplement soil with collected phosphorus fertilizer. Then, as need increased, we began to pull it from the ground, resulting in benefits and problems that will be discussed at the summit. Just as people of the past could not have predicted the tangle of horrible and awesome that is the human condition today, we look ahead and have a hard time discerning what will be worse, what will be better, and how they will be jumbled up together. Phosphorus sustainability impacts everyone, in conditions ranging from cutting-edge scientific laboratories to sustenance farms.

What happens next in the story of phosphorus sustainability is going to depend on how it is told and who is telling it to what audience. Participants in this summit will influence that story, so the words they will choose are important. Their remarks and perspectives may eventually define the messages that will change policy, advance industrial practices, and be broadcast to the public.

At the first exhibition site at the Desert Botanical Garden, glass containers support wooden planting troughs. The troughs are filled with soil and growing things. The glass is filled with sediments, living plants and marine life, images, data, and messages. The glass and old wood forms a 7′ x 7′ x 8′ hollow structure reminiscent of a greenhouse. Light shining through activates living processes and reflects colored shadows that change throughout the day. Viewers can leave their impressions about phosphorus sustainability, in writing or audio documentation.

About the Artists:
Angela Cazel Jahn makes spaces and images for people to play in and think about. Her work is project-based, and often involves viewers as participants. After graduating from Arizona State University with a fine art degree, she veered into the nonprofit field for nearly a decade to serve as co-founding Artistic Director of the Children's Museum of Phoenix. Now she collaborates with individuals, organizations, and institutions to explore questions and illustrate ideas.

Frances McMahon Ward is a visual storyteller working primarily in the areas of video and installation. Her work comes from personal experiences and observations of the world. She has a tendency to latch on to an idea, and when she has thoroughly researched it she begins to construct a narrative, which serves as the backbone for a work. Frances holds a Master's in Fine Arts–Intermedia from Arizona State University. Currently, she holds a position at Xavier College Preparatory, where she teaches courses in the Digital Arts Department and serves as curator/preparator for Xavier's Stark Gallery.

Cory Dunnington grew up in Vancouver and now makes her home in Phoenix. She is a full-time firefighter for the city of Phoenix but has been exercising her creativity by working with kiln-formed glass since 2005. Dunnington enjoys the endless possibilities of melting, manipulating, and molding glass into different shapes and forms. She attended Pilchuck Glass School in 2009 and was awarded a scholarship to Pilchuck in 2010. Her art reflects her passion for color and light patterns as well as the patterns and textures found in nature and in industrial urban life.

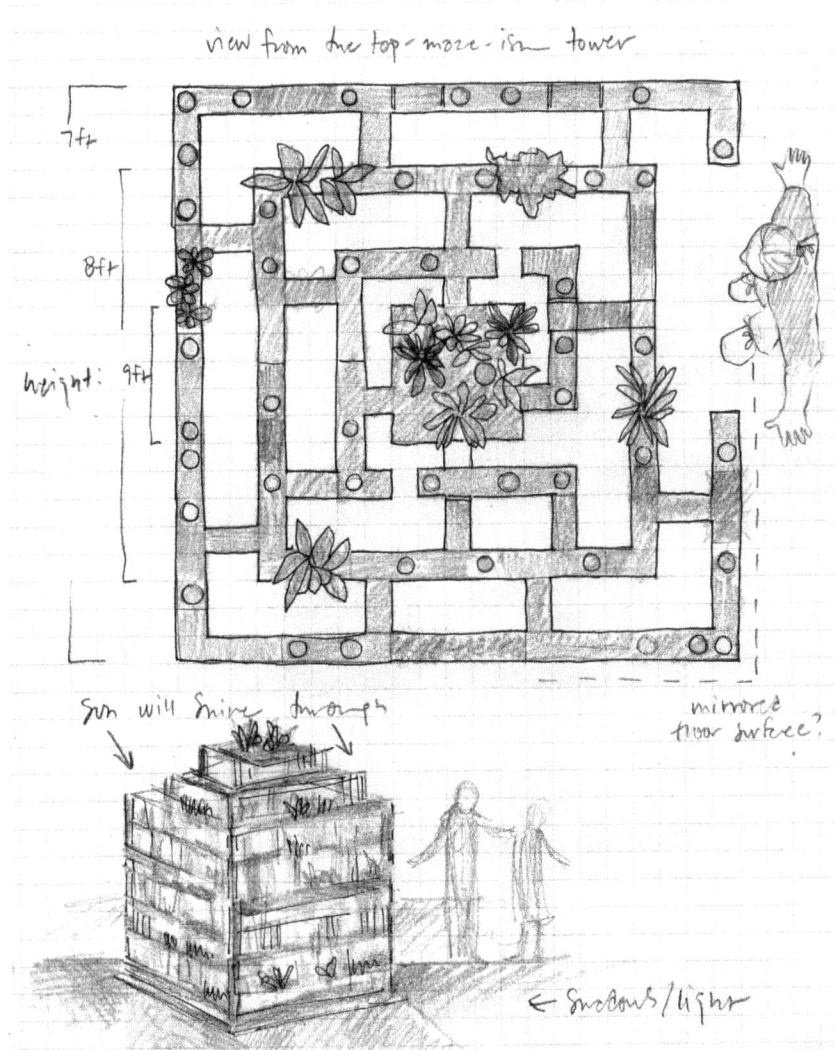

Angela Cazel Jahn, Frances McMahon Ward, and Cory Dunnington; Sketch of: *Our Floating Days*, mixed media sculpture, 8' x 8' x 7', 2011

> **BOX 8.1**
> **CHAPTER 8 OBJECTIVES**
>
> - Describe four research approaches that help scientists better understand the complex manner in which P flows through our societies.
> - Describe how understanding P stocks and flows help us understand where we are wasting P resources.
> - Highlight how some research approaches are evolving to consider temporal, spatial, human, and contextual system factors that affect P-use and flows. Describe a recent research effort to engage system actors such as fertilizer producers, farmers, and policy makers.
> - Present a vision for more holistic research, which integrates such research with farmer education and primary and secondary education, in order in make systems thinking part of everyday life.

INTRODUCTION

The societal systems and agroecosystems through which P flows are complex and require insight into how factors such as farmer livelihood, environmental quality, resource availability, and soil fertility interact as part of a dynamic and structural system of systems. A better understanding of these interactions can result in the development of more robust and sustainable solutions to the P (mis)management problem. Such thinking is starting to gain traction among scientists (Figure 8.1). For example, members of the European Science Foundation (ESF) recently recognized the importance of understanding the complexity of our "Earth system" in order to better support sustainable responses to global change. Indeed, the ESF is calling for scientists to follow an integrated approach from the beginning, involving natural, social, and human sciences and employing transdisciplinary (science-society) research and educative approaches (ESF 2011).

The previous seven chapters of this volume explore various aspects of these complexities in detail, leading us to conclude that more effective governance structures, beyond the conceptual, are needed to reduce the pollution effects of excess phosphorus, while maintaining adequate food security for a growing global population (Elser and Bennett 2011). In this chapter, we illustrate, through a series of four research approaches, how scientists are contributing to a better understanding of the complex dynamic relationships among and between natural and human systems and also promoting sustainable P-use and management.

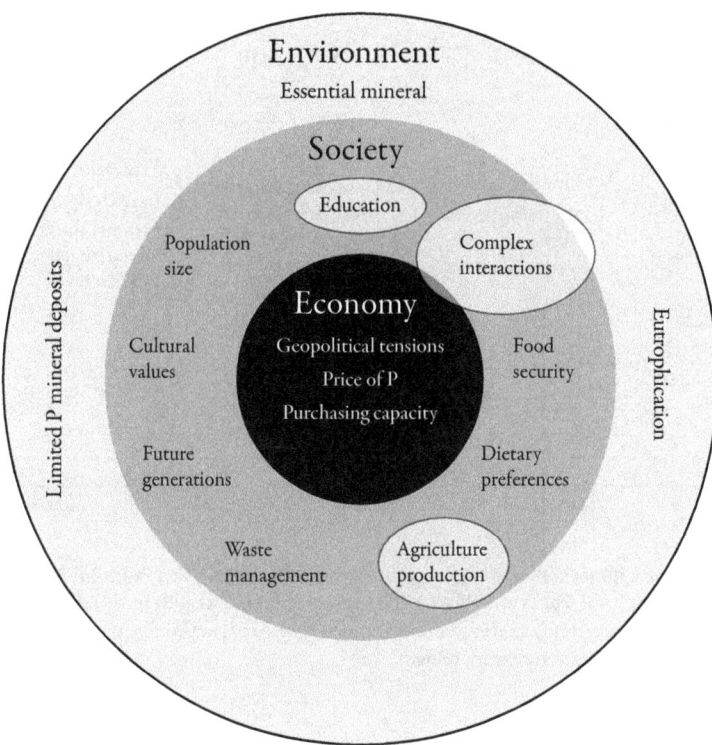

FIGURE 8.1 Our chapter addresses the future of sustainable P in the social sphere, including education efforts, recognizing complex interactions among various global stakeholders, and the demands placed on agricultural systems.

Figure 8.2 depicts the four approaches for studying the **complex systems*** around which this chapter is organized. The chapter will focus on progressively greater temporal, contextual, and human system aspects. **Material Flow Analysis (MFA)*** provides a "snapshot" of a material's movement through a system, revealing P stocks, flows, and losses at a given time. To characterize how interactions between system components and processes can change over time, space, and in response to dynamic contextual factors, some recent research is viewing ecosystems, including agroecosystems (farm systems), as **complex adaptive systems*** (Darnhofer, Fairweather et al. 2010). The end products of many of these studies are models that simulate what a farm system or society will look like under alternative future scenarios. To bring more real-world expertise into their studies, some scientists are developing **transdisciplinary*** and participatory research processes that involve farmers and other system actors. Such cooperative research creates opportunities for mutual science-society, or "social learning," producing more representative scientific data and giving actors opportunities to consider their role in their society (Eshuis and Stuiver 2005; Fairweather and Hunt 2011). Finally, **education reform,*** incorporating complex

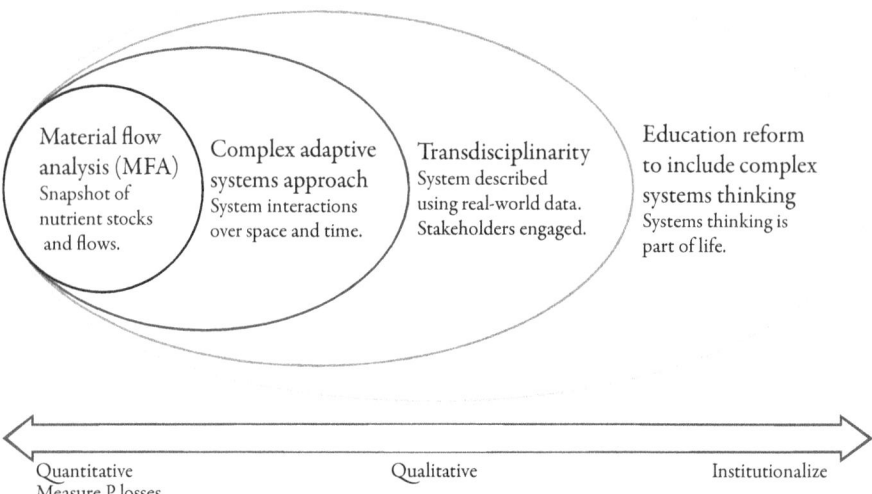

FIGURE 8.2 Four approaches for studying complex systems, starting (left) with MFA, which provides a "snapshot" of P stocks and flows. Each subsequent method adds more temporal, contextual, and human system aspects of societal and agrosystems, improving our understanding of P flows and involving more actors from society.

systems thinking* into farmer extension training, multi-stakeholder collaborations, and primary and secondary education, can help make systems thinking part of the way we think and communicate in our everyday lives (Plate 2010), supporting more enduring transitions toward sustainable P-use and management.

MATERIAL FLOW ANALYSIS

Informed decisions about P require knowledge of where P is stored and how it moves through the environment at different scales. Thus, characterizing P stocks and flows is a key way for science to contribute toward P sustainability solutions. For example, Material/Substance Flow Analysis (MFA/SFA) is a particularly useful tool to support resource policy decisions. MFA is a systematic assessment of the stocks and flows of materials within a system defined in space and time; it connects sources, pathways, and intermediate and final sinks of a given material (Brunner and Rechberger 2004). MFA studies can provide valuable information about the existing amounts of targeted materials and can provide a framework for discussion to improve communication among policy makers. These types of studies can also provide detailed information for public use (Wernick and Frances 2005). Various authors have analyzed P flow from the economical use and recycling perspective (Smil 2000), Neset et al. (2008), Li et al. (2010), Matsubae-Yokoyama et al. (2010).

From these snapshots we might better be able to go beyond the "once-through mode of societal phosphorus metabolism" (Liu 2008).

A sophisticated example of MFA work is a recent study by Matsubae (2011) to understand the virtual (both direct and indirect) P flows in Japan in 2005. The results of the study give an estimate of how much phosphorus ore was required for Japan's agricultural consumption in 2005. We can see in Figure 8.3 the virtual flows of P to (and through) the Japanese economy. Japan consumed 3662 kt of phosphorus ore from overseas countries to produce agricultural products. A total of 6160 kt (3662 + 407 + 240 + 1077 + 774) of ore was required to support Japan's economy annually. Figure 8.3 also shows which countries were suppliers of the 6160 kt.

This study also yields important insight into how much P is lost in processing. Approximately half of the imported phosphate ore (left side of Figure 8.3) was transformed into fertilizer, and used to produce agricultural products, yet only 763 kt of phosphorus ore was eventually "eaten" (consumed); the rest was dissipated in soil and water. This "eaten" phosphate ore was 20 percent (763/3743) of the imported virtual phosphorus used for fertilizer, food, and feed in Japan. Now that the MFA has revealed the large difference between "eaten" and virtual phosphorus, the question becomes: how can we reduce the P losses? Japan currently follows an intensive program of reusing plant residues and recycling of livestock and human excrements, so what more can be done? The greater volume of the imported agricultural products and the less the phosphorus yield food consumption, the lower the ratio of "eaten" to imported phosphorus. Here phosphorus yield loss means the difference between

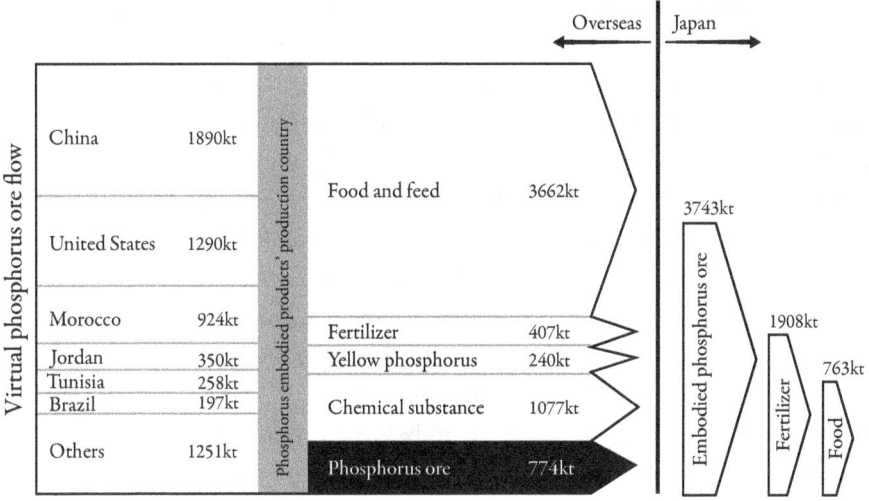

FIGURE 8.3 Virtual phosphorus ore requirement of Japanese economy in 2005 (Matsubae, Kajiyama et al. 2011).

phosphorus input and output such as P in fertilizer and crops, feed and meat, and feed and cultured fish. Phosphorus yield loss cannot be avoided in agricultural production, but can be partially reduced by soil survey and control of fertilizer input. However, production of meat requires more phosphorus than production of beans, and sometimes farmers use too much fertilizer for high-grade products. These two facts lead to the conclusion that the economic enrichment and change in food preference from plant to animal proteins require more and more phosphorus to produce our daily food. In cases where the preference is for more phosphorus-intensive agricultural products, such as meat, it might lead to a smaller "eaten" P / virtual P ratio, due to the greater phosphorus yield loss. Additionally, an excess supply of phosphorus and the promotion of phosphorus accumulation in soil due to ineffective P-use might lead to a smaller "eaten" P / virtual P ratio (Vance, 2001).

RECOGNIZING AGROECOSYSTEMS AS "COMPLEX ADAPTIVE SYSTEMS"

How can scientists and stakeholders better understand how P flows through a system over space and time? MFAs can give many countries a rough snapshot of P flows, but they do not predict the future. Participants in the July 2011 European Scientific Workshop, "Designing Phosphorus Cycling at the Country Scale," held in Bordeaux, France, expressed concerns about how MFAs do not show shifts between plant-available and non-available P. They recommended more integration between MFA and other approaches that, for example, simulate P movement in soil over time and space (Cors 2011).

Complex adaptive systems research theory (second component of Figure 8.2) goes beyond the classical research view of farms to bring in temporal, spatial, contextual, and human elements of a system as well as the natural environment. This approach calls for sustainability scientists to pay more attention to individual and societal drivers that influence P-use, and to the consequences of P management practices.

In a classical research approach (see Table 8.1), farm systems are viewed not as dynamic systems (Belcher 2004, Schlecht and Hiernaux 2004) but are typically described as being in a steady state, with nutrient and cash flow balances expressed as statistics (amounts or rates). A complex adaptive systems view introduces the idea that the rate and nature of change in farm processes, as well as the interactions between system elements, are dynamic. For example, the change in livestock population could shift from a cyclical pattern, which depends upon annual births and deaths, to an erratic pattern that follows market prices for meat, or to a pattern of decline that reflects a decline in forage food due to climate change. Here the investigative lens is the co-evolution of humans

TABLE 8.1

COMPARISON OF RESEARCH PARADIGMS: CLASSICAL FARMING SYSTEMS VIEW AND COMPLEX ADAPTIVE SYSTEMS THEORY APPROACH (ADAPTED FROM DARNHOFER, BELLON ET AL. 2010)

	Classical farming systems view	Complex adaptive systems theory
Underlying idea	General systems theory; show nutrient and cash balances.	Co-evolution of humans and nature; systems adapt; interactions are focus.
Time	Atemporal but some linear projection into the future; no change in the system's dynamics.	Time is a key variable: behaviors evolve and reflect history, dependence on other system dimensions, and irreversibility of some pathways.
Dynamics of system	Static approach, steady-state, equilibrium view.	Perpetual disequilibrium: rate and nature of interactions change over time and space.
Context	Contextual factors such as climate, market forces, and population are assumed to be steady over time. Focus on farmer perception and agricultural sector.	Context is constantly changing according to strength, place, timing, and direction of influence. Given systems interactions, there is a need to include all influential sectors.
Performance indicators	Production-focused: crop yield; economic returns to the farm.	Sustainability-focused: soil fertility (nutrient balances), farmer livelihood (including social assets), farmer flexibility.
Farmer strategies	Operator (farmer) focus. Efficiency and a large workforce promote success.	Focus on farmer as part of local and regional communities; livelihood is linked to resilience; flexibility, learning, network-building promote success.

and nature, for which several approaches have been developed (Matthews and Selman 2006, Scholz 2011).

Generally, classical studies do not introduce a temporal element and/or neglect to examine nonlinear behavior in a farming system over time and space. In complex adaptive systems thinking, time is a key variable, and studies of P nutrient use should consider history and path-dependent processes. For example, Vietnam's ever-growing demand for national and export food supplies has created an irreversible trend for farmers using inorganic fertilizer, a product they are now locked into using to meet market demands (Bo and Mutert et al 2003). It is important to recognize dynamic interactions at different hierarchical levels of the system, from global influences like climate change, to soil-crop-livestock interactions, to P soil dynamics (which affect how available P is to crops).

Many classical studies see the agricultural sector as the context and involve appropriate actors, such as amendment suppliers or food distributors, who are most often described through the eyes of the farmer. From a complex adaptive systems perspective, the environmental and socioeconomics contexts in which farm systems operate are constantly changing in the strength, timing, and direction that they influence the farm system. Thus, it is important to investigate the influence from actors such as policy makers, farmers, and fertilizer processors by involving them in research through participatory and transdisciplinary approaches.

Classical studies also measure performance of farm systems based on crop yield, in which the main actor of interest is the farmer (operator), who achieves success through the efforts of a large workforce and/or through efficiencies in work processes. A complex adaptive systems approach also tracks performance indicators such as soil fertility, farmer livelihood, farmer flexibility, and returns to land and labor related to sustainability. The farmer is viewed as part of local and regional communities, where flexibility, ability to learn, and network building promote success. Recent work by Darnhofer, Bellon et al. (2010) describes how "Change is then no longer seen as a disturbance, but as a trigger for the reorganisation of resources, and for the renewal of the farm organisation and activities. Implementing these strategies comes at a cost, so that farmers need to tackle the inevitable trade-offs between efficiency and adaptability. However, unless farmers master this challenge they cannot ensure the sustainability of their farms."

TRANSDISCIPLINARITY: INCORPORATING MORE REAL-WORLD DATA

Involving multiple stakeholders in natural resource management efforts is fundamental to understanding and improving P management in farm systems, where it is estimated that about half of P is lost (Cordell, Drangert et al. 2009). The benefits of participatory and transdisciplinary research (see third component of Figure 8–2) are becoming obvious to more and more researchers: 1) Stakeholders have unique expertise about their situation and tapping into their knowledge often allows a study to produce a more accurate characterization of an agroecosystem (Probst and Hagmann 2003; Roder 2004; Scholz 2011) and 2) Engaging stakeholders in studies, programs, and policy development makes it more likely they will embrace and follow the new programs (Cors 2004; Eagan and Cors 2003; Nastasi and Varjas et al. 2002; Pomeroy and Douvere 2008). The question then becomes, how can real-world data—information from farmers, credit agencies, extension services, fertilizer companies, and so on—be brought into research about agricultural systems?

One example of groundbreaking work in science-society cooperation to study fair management approaches, technologies, and policy options that promote sustainable P management is the Global Transdisciplinary Processes for Sustainable Phosphorus Management or Global TraPs project (Ulrich and Cors 2011). Global TraPs is an international transdisciplinary project that involves a diverse group of participants representing science, industry, NGOs, and other stakeholders in processes of joint problem definition, mutual learning, and capacity building. Please see chapter 10 for the Global TraPs URL for an example of this project. A transdisciplinary process is designed to apply the unique knowledge of individual scientists and stakeholders to problem solving, through collaboration, consensus, and integration. The idea is to engage stakeholders and decision makers in formulating alternatives for policies and programs they can eventually implement. Done properly, transdisciplinary processes should also produce relevant, socially robust orientations that contribute to generation of new scientific knowledge and theory building.

The purpose of the Global TraPs project is to develop a better understanding of the opportunities and risks linked to how humans use P. The project is guided by joint leadership from ETH Zürich's Natural Social Science Interface (NSSI) and the International Fertilizer Development Center (IFDC). Global TraPs integrates knowledge from case studies organized along the P supply chain in node groups, from the "Exploration" to the "Recycling" node. Two projects in the "Use" node are examining how farmers evaluate trade-offs between farm productivity and farm resilience and how to improve the adaptive capacity of farm systems.

Other international efforts to promote sustainable P management that have recently been established include Arizona State University's Sustainable Phosphorus Initiative (SPI) and the Global Phosphorus Network (GPN).

EDUCATION REFORM: INSTITUTIONALIZING COMPLEX SYSTEMS THINKING

The fourth component illustrated in Figure 8.2, Education Reform, refers to a broader, more visionary perspective of how scientists could become agents of change—promoting systems thinking among system actors. This transition involves reforming education of farmers, primary and secondary students, and the general public to incorporate complex systems thinking.

Often the **epistemic*** barriers for humans to move toward sustainable practices come from a "psychological distance" from environmental problems (Carolan 2006). For example, humans cannot taste, touch, hear, feel, or see PCB, nor did they consider in the 1940s how widespread use of high-yield hybrid corn would, over the long term, reduce genetic variability and thus increase vulnerability of the crop to disease (Pringle 2003). Similarly, research has shown that scientists also find it challenging to think about the important dimensions of environmental problems and, for example, to incorporate concepts of scale into sustainability research (Gibson, Ostrom et al. 2000). Thus, complex systems thinking research must be embraced by more scientists and universities and brought into technical training and public education systems worldwide. Bringing complex systems thinking into farmer training and primary and secondary school learning in both developing and developed countries will likely be important for achieving sustainable P management in the future.

From a complex adaptive systems perspective, farmers who can learn, who are flexible, and who have developed a larger number of options for managing their farms adapt more easily to changing conditions (Darnhofer, Fairweather et al. 2010). However, barriers to understanding farm systems as complex adaptive systems, and developing and taking action to embrace sustainable transitions, lie not only with farmers (Carolan 2006). A better understanding of these barriers can only be gained by looking at all relevant actors, at both individual and societal levels, and at how they interact with one another. As such, promoting complex system thinking and pro-environmental behaviors regarding farm systems means involving not only the farmer, but also the communities in which they live and work and the people they feed. So, for example, collective mapping of P nutrient cycling through a farm system should involve not only farmers, but also policy makers, scientists, extension agents, fertilizer producers, food distributors, and the community at large. Such integrated, transdisciplinary education approaches present exciting opportunities to aggregate and apply knowledge about local and regional P management. Developing such knowledge is a first step in exploring policy options and may also reveal traditional P management approaches that can serve as best practices.

To reach the decision makers of tomorrow we need to reform primary and secondary education policy today. Only recently have studies recognized that systems thinking is not intuitive and that both children and adults commonly approach system problems with open-loop and one-way causal thinking (Booth Sweeney and Sterman 2007; Frischknecht-Tobler, Nagel et al. 2008). It is of course promising to see that research also suggests that "students receiving systems-oriented instruction exhibit significantly increased ability to understand information about a complex environmental system, despite having received no specific training in natural resources" (Plate 2010). More research should be integrated with teaching and learning about complex systems in primary and secondary school classrooms. Much of this work needs to happen in the field, where scientists employ action research strategies to enable school administrators and teachers to assess the degree to which complex adaptive systems thinking is practiced in the classroom and on the playground.

Education reform to incorporate complex adaptive systems thinking into farmer and citizen education can support these multi-stakeholder efforts and, at its best, would promote systems thinking in society. Imagine a world where fertilizer producers, farmers, grocery store owners, parents, and scientists are all equipped with complex systems thinking skills to discuss how P is used and managed in their society.

CONCLUSION

If humans are to more sustainably use and manage P, it is essential that we better understand how P flows through Earth's systems, particularly our agricultural and urban systems, where human activity has most greatly affected P cycling. Researchers are developing sophisticated Material Flow Analysis methods to characterize P flows and, for example, to compare demand and supply of P resources (Matsubae et al. 2010).

To assist in development of P management programs that will endure over time and be locally relevant, it is critical that research also characterize the complex, dynamic relationships among and between natural and human systems. To this end, some researchers are already incorporating system factors such as time, space, and system contexts into their research, developing frameworks such as complex adaptive systems theory. By engaging actors, such as farmers, fertilizer producers, credit lenders, and policy makers into the transdisciplinary research process, they can more accurately characterize P flows and also support acceptance and better implementation of sustainable P-use policy and programs.

Engaging actors in such complex systems research introduces them to new opportunities to better understand and influence their world. Echoing the Chinese proverb, *Give a man a fish and you feed him for a day. Teach a man to fish and you feed him for a lifetime*, we believe that actors can become an even bigger part of developing solutions for sustainable P management, and that education is the key. By integrating research with education—farmer extension, consumer education, and in our schools—scientists can help institutionalization complex systems thinking and ideas about sustainable P cycling into everyday work and life. What if we all—from fertilizer traders to farmers to first-graders—had the skills and opportunities to talk with each other about how P travels through our part of the world?

BOX 8.2
CHAPTER 8 SUMMARY

Visions for enduring transitions toward sustainable P-use and management:

- We have reviewed a series of four increasingly holistic approaches for studying the complex nature of P flows through Earth's societies.
- *Material Flow Analysis* enables scientists to characterize P stocks, flows, and losses through regional-scale systems (nations, watersheds, etc.) at a given point in time.
- A *complex adaptive systems* approach allows scientists to characterize how P flows change over time, space, and in response to contextual factors, including human and climate factors.
- Through *transdisciplinary research and process* scientists involve farmers, fertilizer processors, and other system actors in developing more real-world data to support studies of P and in promoting sustainable policy and practice.
- By incorporating complex systems thinking into farmer extension training, transdisciplinary research, and primary and secondary *education*, scientists can promote institutionalization of sustainable P-use practices.

We envision a world where people, from fertilizer processors to farmers, are equipped with complex systems thinking skills to discuss how P is used and managed in their society.

REFERENCES

Belcher, K. W., M. M. Boehm, M. E. Fulton 2004. Agroecosystem sustainability: a system simulation model approach. *Agricultural Systems* 79:225–241.

Bo, N. V., E. Mutert, and C. D. Sat. 2003. *Balanced Fertilization for Better Crops in Vietnam (BALCROP): Potash & Phosphate Institute of Canada* (Southeast Asia Programs).

Booth Sweeney, L., and J. D. Sterman. 2007. Thinking about systems: student and teacher conceptions of natural and social systems. *System Dynamics Review* 23(2/3):285–312.

Brunner, P. H., and H. Rechberger. 2004. *Practical Handbook of Material Flow Analysis*. Lewis Publishers.

Carolan, M. 2006. Do You See What I See? Examining the epistemic barriers to sustainable agriculture. *Rural Sociology* 71(2):232–260.

Cordell, D. A., J. O. Drangert, and S. White. 2009. The story of phosphorus: Global food security and food for thought. *Global Environmental Change* 19:292–305.

Cors, R. 2011. *Notes from the European scientific workshop on Designing phosphorus cycle at country scale*. Bordeaux, France, ETH Zürich, Natural and Social Science Interface.

Cors, Rebecca. 2004. *Improving U.S. State Agency Environmental Management through Organizational Learning, Action Research, and International Exchange*. Gaylord Nelson Institute for Environmental Studies, Land Resources Program Thesis, University of Wisconsin—Madison.

Darnhofer, I., S. Bellon, B. Dedieu, and R. Milestad. 2010. Adaptiveness to enhance the sustainability of farming systems. *A review, Agronomy for Sustainable Development*. 30:545–555.

Darnhofer, I., J. Fairweather, and H. Moller. 2010. Assessing a farm's sustainability: insights from resilience thinking. *International Journal of Agricultural Sustainability* 8(3):186–198.

Eagan, P. and R. Cors. 2003. *An Organizational Change Strategy to Improve Environmental Protection*. International Symposium on Electronics and the Environment in conjunction with the IEEE Electronics and Recycling Summit Proceedings, Boston, MA.

Elser, J., and E. Bennett, E. 2011. Phosphorus cycle: A broken biogeochemical cycle. *Nature* 478:39–31.

ESF (2011). *Presentation: ESF-COST "Frontier of Science" Initiative & ESF Forward Look*. European Response to Grand Challenges in Sustainability Research, Brussels, Belgium.

Fairweather, J., and L. Hunt. 2011. Can farmers map their farm system? Causal mapping and the sustainability of sheep/beef farms in New Zealand. *Agriculture and Human Values* 28:55–66.

Eshuis, J., and M. Stuiver. 2005. Learning in context through conflict and alignment: Farmers and scientists in search of sustainable agriculture. *Agriculture and Human Values* 22:137–148.

Frischknecht-Tobler, U., U. Nagel, and H. Seybold. 2008. *Systemdenken: Wie Kinder und Jugendliche komplexe Systeme verstehen lernen*. Bern, Switzerland, Stämpfli Publikationen.

Gibson, C. G., E. Ostrom, and T. K. Ahn. 2000. The concept of scale and the human dimensions of global change: a survey. *Ecological Economics* 32:217–239.

Li, S., Z. Yuan, J. Bi, and H. Wu. 2010. Anthropogenic phosphorus flow analysis of Hefei City, China. *Science of the Total Environment* 408(23):5715–5722.

Liu, Y., G. Villalba, R. U. Ayres, and H. Schroder. 2008. Global phosphorus flows and environmental impacts from a consumption perspective. *Journal of Industrial Ecology* 12(2):229–247.

Matsubae, K., J. Kajiyama, T. Hiraki, and T. Nagasaka. 2011. Virtual phosphorus ore requirement of Japanese economy. *Chemosphere* 84(6):767–772.

Matsubae, K., J. Kajiyama, and T. Nagasaka. 2010. WIO-MFA on Phosphorus Recovery from the Waste Materials. EcoBalance 2010, The 9th International Conference on EcoBalance November 9–12, Tokyo.

Matsubae-Yokoyama, K., H. Kubo, K. Nakajima, and T. Nagasaka. 2009. A Material Flow Analysis of Phosphorus in Japan. *Journal of Industrial Ecology* 13(5):687–705.

Matthews, R., and P. Selman 2006. Landscape as a focus for integrating human and environmental processes. *Journal of Agricultural Economics* 57(2):199–212.

Nastasi, B. K., K. Varjas, S. Schensul, K. T. Silva, J. J. Schensul, and P. Ratnayake. 2002. The Participatory Intervention Model: A framework for conceptualizing and promoting intervention acceptability. *School Psychology Quarterly* 15(2):207–232.

Neset, T. S. S., H. P. Bader, R. Scheidegger, and U. Lohm. 2008. The flow of phosphorus in food production and consumption—Linkoping, Sweden, 1870-2000. *Science of the Total Environment* 396(2–3):111–120.

Plate, R. 2010. Assessing individuals' understanding of nonlinear causal structures in complex systems. *System Dynamics Review* 26(1):19–33.

Pomeroy, R., and F. Douvere. 2008. The engagement of stakeholders in the marine spatial planning process. *Marine Policy* 2(5):816–822.

Pringle, P. 2003. *Mendel to Monsanto—The Promises and Perils of the Biotech Harvest*. Simon & Schuster, New York,.

Probst, K., and J. Hagmann. 2003. *Understanding participatory research in the context of natural resource management—paradigms, approaches and typologies*. Agricultural Research & Extension Network, Network Paper No. 130.

Roder, W. 2004. Are mountain farmers slow to adopt new technologies? Factors influencing acceptance in Bhutan. *Mountain Research and Development* 24:114–118.

Schlecht, E., and P. Hiernaux. 2004. Beyond adding up inputs and outputs: process assessment and upscaling in modelling nutrient flows. *Nutrient Cycling in Agroecosystems* 70(3):303–319.

Scholz, R. W., Ed. 2011. *Environmental Literacy in Science and Society: From Knowledge to Decisions*. Cambridge University Press, Cambridge, UK.

Smil, V. (2000). Phosphorus in the environment: Natural flows and human interferences. *Annual Review of Energy and the Environment* 25:53–88.

Ulrich, A.E., and Cors, R. 2011. Global TraPs: An international effort to promote sustainable use of phosphorus. In: *Ecoregion Perspectives. Sustainable Agriculture in the Baltic Sea Region in times of peak phosphorus and global change*. Baltic 21 Series No. 4/2011:26–28.

Vance, C.P., 2001. Symbiotic nitrogen fixation and phosphorus acquisition. Plant nutrition in a world of declining renewable resources. *Plant Physiology*. 127: 390–397.

Wernick, I. K., and H. Frances 2005. Material flow accounts: A tool for making environmental policy, *World Resource Institute database:* Metal & Mineral Flows.

9

Future Scenarios for the Sustainable Use of Global Phosphorus Resources

P IS FOR PREFERRED (P)FUTURES

Daniel L. Childers, Zachary Caple, Cynthia Carlielle-Marquet, Dana Cordell, Vanda Gerhart, David Iwaniec, Stuart White

Todd Daniel Grossman; Detail of: *P is for Pail, P is for Preservation,* and *P is for Phosphorus (three-piece series)*; Watercolor, 4" x 6," 2011

Todd Daniel Grossman
P is for Pail; P is for Preservation; P is for Phosphorus
(three-piece series)

Scientific Collaborator:
Robert Mikkelsen, Director, International Plant Nutrition Institute

Description of Artwork:
When I was working on the concept for this piece, I asked Dr. Mikkelsen what image stood out for him when he thought about phosphorus sustainability. He said, simply and beautifully: "A farmer in Africa with extremely P-deficient corn plants... He was struggling to grow sufficient food to provide for himself and his family. A bucket of P fertilizer would drastically change their lives. The simple things we take for granted can be a matter of life or death elsewhere." This series of three paintings represents the physical aspect of phosphorus (P is for Pail), the human need for phosphorus (P is for Preservation), and the chemically finite aspect of phosphorus (P is for Phosphorus).

About the Artist:
Todd Daniel Grossman is a painter specializing in miniature drybrush watercolor works. A graduate of Pomona College, with a degree in English literature, he is a classically trained pianist, composer, and lyricist, currently finishing an original opera. He and his husband of 14 years live in a historic Phoenix 1920s Craftsman. A published professional photographer, he also teaches piano, voice, art, and gymnastics. In his spare time, he knits.

BOX 9.1
CHAPTER 9 OBJECTIVES

- Articulate a "way forward" for managing phosphorus that is grounded in sustainability science theory and practice that includes:
  - Identify key stakeholders who should participate in the process.
  - Implement collaborative visioning of sustainable future states.
  - Use backcasting scenarios to determine the transitions and interventions necessary to get from the current state to envisioned sustainable future states.
  - Acknowledge the importance of multiple paths forward and the need for nimbleness and flexibility to accommodate known and unforeseen changes in exogenous drivers and internal system dynamics.
  - Identify policy challenges and general strategies that may be necessary to overcome the institutional and infrastructural inertia of the current system.

INTRODUCTION

The sustainability paradigm is changing how we think about the future. Whether it is with adjustments to efficiency and consumption, with spectacular technological advancement, or with transformational social movements, a variety of perspectives on how to address sustainability challenges are populating a global dialogue on our long-term future. While potential pathways to sustainability are filled with complex near-term problems, some of which will emerge unexpectedly, sustainability challenges also require long-term and holistic perspectives (*see* **sustainable scenario development***). In this chapter we articulate a framework to envision and develop **backcasting*** sustainability solutions for the global P challenge, and we provide an overview of potential transitional strategies and interventions. We summarize a range of measures, or actions, that would result in reduced impact on phosphorus resources and less impact of phosphorus use. Finally, we identify a range of barriers to the implementation of these measures, including past reliance on **traditional forecasting,*** and then discuss generic policy tools to help overcome these barriers.

STAKEHOLDER IDENTIFICATION

The co-production of sustainable solutions and strategies for achieving those solutions may be greatly improved by effectively engaging all relevant stakeholders (Arnstein 1969; Fischer 1993; Clark et al. 2006; van Kerkhoff and Lebel 2006; Guston 2008; Robinson 2011; Talwar et al. 2011). We define "**stakeholder***" as any person or group who can affect or is affected by a particular situation or system. The global phosphorus (P) sustainability challenge may, in fact, have nearly 7 billion human stakeholders, because everybody needs food and we cannot feed ourselves without a reliable supply of P. It could be argued that this number of stakeholders makes stakeholder engagement difficult at best. However, there are some key stakeholder groups that should be considered in the development of a plan for sustainable solutions. A cursory identification of them includes: phosphate mining companies, fertilizer manufacturers and distributors, agri-business, farmers, food producers, food distributors and retailers, dieticians and nutritionists, consumers and homeowners, wastewater managers, natural resource managers, policy makers, scientists and researchers, and entrepreneurs. Global stakeholder engagement appears challenging, at least initially, but local or regional engagement of most of these groups should be part of implementing sustainability solutions.

BACKCASTING TOWARD A SUSTAINABLE FUTURE

In sustainability science parlance, the next step after **visioning*** sustainable future states is to determine how best to "get there" from the current situation. One approach is to develop scenarios that begin at the present and project toward desirable and possible future states. These forecasted scenarios start from the current or past state and follow trajectories that are largely driven by the current state of the system or situation. Forecasts are representations of future outcomes that are determined from the current state or a historical reference that is then advanced forward through time, resulting in outcomes that are foremost determined by past and current conditions. Past trends (path-dependent conditions) are often important in determining the future. Near term, predictable futures that are highly dependent on past or current conditions can be best extrapolated using forecast-based approaches.

A limitation of forecast approaches is that they project the problems of today onto a future that is both nonlinear and subject to the limits of human creativity. If we are interested in long time horizons, anticipate a large number of surprises, or expect or require transformative changes, then we need different future-oriented approaches. Path-dependent forecasting may not facilitate transformative thinking or encourage transformative change (i.e., "thinking about a whole new box" rather than simply "thinking outside of the box"). Another approach is to start by crafting a desirable future state and then analyzing potential pathways from this vision back to the present-day situation. This is known as backcasting (Figure 9.1; Robinson 1982; Dreborg 1996; Robinson 2000, 2003). While backcasting is often more challenging than path-dependent scenario development, it has the advantage of highlighting where transformative changes may be needed. Some alternative futures may require radical changes and yet-to-be-imagined technological and social innovations (Höjer 2000). Since backcasting is based on desirable futures and not locked into current technologies or mind-sets, its relevance may well increase as radical and transformative changes become more critical. Put simply, if we cannot easily get from a sustainable future state to our current, presumably unsustainable situation, then simple

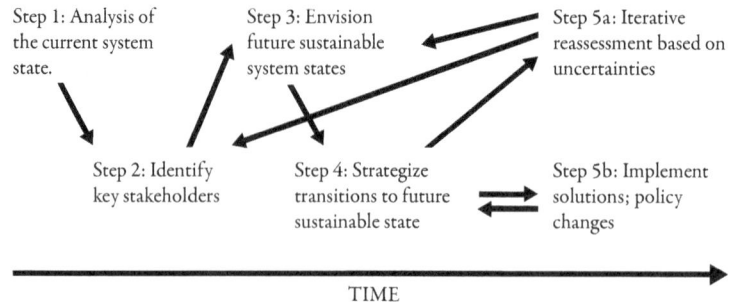

FIGURE 9.1 A conceptualized timeline for "moving forward" with sustainability solutions to the global phosphorus challenge. The two steps on the left are components that are covered in this book, while the four steps on the right are covered in this chapter. Notably, this generalized road map should be applicable to most sustainability challenges.

"tweaking" of the current system will likely be insufficient to get to that desirable future state. At some level, transformational change will be necessary. That said, we posit that forecasting and backcasting approaches are complementary techniques and both are essential approaches to future thinking.

BOX 9.2
A SHARED VISION OF A SUSTAINABLE PHOSPHORUS FUTURE

Our future scenarios envision increased globalization of knowledge production, technology transfer, and environmental stewardship but an economy driven more by local and regional entrepreneurial innovation and less by global market forces. One future vision might include global governance by an international policymaking body that ensures the equitable use of all critical non-renewable global resources. This international governance body might have separate, smaller committees tasked with the global management of specific resources in the context of cross-resource interactions and trade-offs, and one such committee would be responsible for setting global policy on P management (e.g., an Intergovernmental Panel for Phosphorus Security, or IPPS; see chapter 8). Such an IPPS would also support international research that drives continued sustainability-focused innovation of the human aspects of the P cycle, from reserve identification and mining to consumption and waste management. Not serendipitously, our envisioned sustainable future will be characterized by a dramatically reduced gap between the rich and the poor, both among nations and among people within a given nation. Much of this economic equalization will be achieved with a major commitment to education that includes focus on a shared human future and an equitable division of critical resources.

Our sustainable future will include largely regional and local approaches to environmental stewardship in the context of globally responsible environmental ethics. People, societies, and nations will be responsible for their environmental footprints, including consumption based on need, not "want," and wastes being treated as resources and recycled accordingly. New attitudes about food will prevail, with most people eating low on the food chain and meat consumption viewed as an occasional option. Sustainability science and solutions both entail systems thinking, and from this holistic perspective we can easily envision collateral benefits to other ecosystem services (and to human health and well-being). Although our vision may outwardly seem to homogenize our societies and even the biosphere, it will fully embrace the rich heterogeneity of values, cultures, and ecosystems that characterize the planet. Our vision of P sustainability will not be one of a few uniform scenarios (e.g., the technogarden or adaptive mosaic scenarios of the Millennium Ecosystem Assessment, 2005). This future will require highly participatory, or bottom-up, governance structures focused on anticipating problems rather than responding to them once the damage has been done (e.g., preventing fires rather than putting them out).

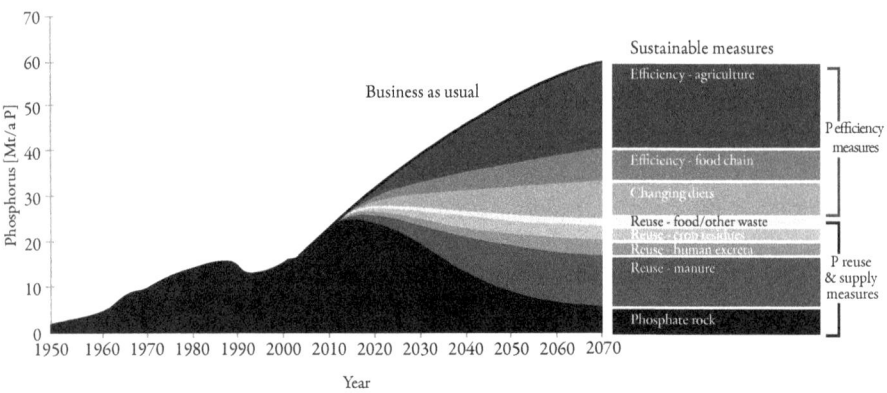

FIGURE 9.2 A preferred sustainable phosphorus scenario for achieving global food security. The sustainability measures on the right are estimated reductions in P-use from a BAU projection and other sources of P from reuse within the system that are likely to ensure sustainable P-use by 2070 with a minimum of "new" P required from mineral P sources (from Cordell et al. 2009).

As with the visioning process we described previously and exemplified in Box 9.2, backcasting exercises should ideally be conducted with stakeholder engagement to the fullest extent possible (Robinson 2003). With regard to P sustainability, backcasting could involve developing strategies that achieve P and food security by reducing the need for "new" mined P relative to a "business as usual" (BAU) future scenario. The BAU scenario is based on currently accepted projections for human population growth, growth in affluence, and dietary demands (Figure 9.2). This simple backcasting exercise is only temporal and has no spatial component to it and thus must be based on globally averaged numbers from budgeting exercises). An important component of this goal is to meet future phosphorus demand while also meeting the world's food needs. Here, backcasting suggests that we may get from a BAU future to a sustainable future by increased efficiency of phosphorus use (P Efficiency Measures in Figure 9.2) and increased recycling and reuse of the P already in the human phosphorus cycle (P Reuse and Supply Measures in Figure 9.2; Cordell et al. 2009). Both involve what Childers et al. (2011) referred to as "closing the human P cycle." In both cases, the goal is to reduce the need for "new" mined phosphate into the future relative to the BAU scenario. Cordell et al. (2009) argued that future food demands cannot be met without an integrated approach that involves substantial increases in efficiency and recycling from mine to field to fork, which are described throughout this book. These efforts include increasing efficiency during mining, fertilizer production and trade, agriculture and food production and processing and consumption; and efficiently recovering P from "waste" streams, such as mine waste, crop and food waste, and manure and excreta (see chapters 3, 4, 5, and 6 for details).

Once vision-based backcasting scenarios or transitions have been formulated, the next step is determining the intervention points (Figure 9.1). These **Interventions*** are, quite simply, "purposeful action by an agent to create change" (Midgley 2003). There may be multiple paths with multiple intervention points, and they are likely to be non-deterministic (that is, not path-dependent) and nonlinear (Loorbach and Rotmans 2006). Solutions associated with intervention points must be integrated, such that no single solution—however transformative—will be adequate. Similarly, solutions must operate across different scales such that they are based on local or regional context but respond to national or global challenges. The P sustainability challenge is far too complex to be left solely to market-based or agriculturally based solutions (Cordell et al. 2010; Childers et al. 2011). Environmental challenges should be addressed with solutions that go beyond cleaning up eutrophied water bodies to integrated approaches focused on the sources of the P challenge. Solutions to the current fragmentation, even non-existence, of institutions to manage P and most other non-renewable resources must be sought. Similar gaps in our knowledge about the entire human P cycle must be filled with interdisciplinary and transdisciplinary teams focused on solutions-oriented environmental, social, and economic research. And finally, many of the transformative solutions that we will require will not be possible until individuals and society recognize that institutional and infrastructural inertias must be overcome. Education is critical to this recognition.

PREPARING FOR AND ADAPTING TO UNCERTAINTIES WITH AN ITERATIVE PROCESS

The process we describe above and map in Figure 9.1 is focused on the future. Yet there are many uncertainties about the future that present challenges to this process. These uncertainties include unanticipated changes in key exogenous drivers that control the system in question, as well as unexpected changes within the system itself. For example, a disruption to global P supply through political instability could cause rapid increases in fertilizer price, which would impact a range of poor farmers and also cause food shortages and food riots such as occurred in 2007 and 2008. An example of an uncertain exogenous driver is sea level rise, which is an important environmental control on and regulator of coastal community policy and management decisions. The most recent predictions from the International Panel on Climate Change (IPCC 2011) predict a meter or more of sea level rise by 2100. However, these predictions are conservative in that they do not account for the instability of the Greenland ice sheet and the West Antarctic ice sheet. If either of these continental ice masses breaks away and slides into the ocean, global sea levels could rise by 6–7 meters in a remarkably short period of time. Envisioning sustainable futures for coastal populations should be flexible enough to account for this possible future change. Changes in political leadership from election to election, and the ensuing instability in policy that these changes often produce, represent internal system dynamics that cannot be predicted, and our solutions-generating process must be nimble enough to adapt to them. At any time, either internal changes or external drivers may alter the decision-making environment enough that sustainable future states should be reenvisioned, that backcasting transition trajectories should be remapped, that intervention points should be re-identified, that the stakeholder groups must be re-formed, or that any combination of these must be revisited. This flexibility and nimbleness comes from ensuring that the process we outline above and in Figure 9.1 is cyclical rather than linear. Not only is there likely to not be a single "correct" vision for a sustainable future, but also these visions must be adaptable to surprises and change from within the system and beyond. In short, getting from our current system state to a sustainable future state requires constant vigilance, attention, and input of energy and new information.

THE CHALLENGES OF SCALE

The future-looking process we describe above is firmly grounded in sustainability science theory (Olson 1995; Costanza 2000; Raskin et al. 2002; Swart et al. 2004; Brewer 2007; Robinson 2011) and has been "field-tested" in several situations, including in a recent application of the technique in the city of Phoenix, Arizona, and in Australia (Cordell and White 2009). As such, the literature reflects success stories that confirm the efficacy of the process at local and sometimes regional scales. The P sustainability challenge is, in many ways, a long-term and global challenge, though. We must think at much larger spatial and temporal scales for solutions. In many ways it may be difficult to envision how the process we describe here might articulate at global or even national scales. The key challenge is likely Step No. 2 in Figure 9.1, where key stakeholders are identified. The larger the scale of a problem (either temporal or spatial scale), the more stakeholders will be involved. It is clear that the process will become overly cumbersome if it remains bottom-up, or driven by large groups of diverse stakeholders, when large-scale solutions are being sought. To that end, it will likely be necessary to take more of a "democratic republic" approach to stakeholder representation for national and global-scale solutions. If care is taken to ensure solid communication between stakeholder representatives and their individual stakeholder constituents, and care is taken to ensure that a representative of all key stakeholder groups is at the table, this process should be successful. None of this can happen without solid leadership, though, which provokes the question: For the P sustainability challenge, where will this leadership come from at either the national or global scales?

MEASURES: EXAMPLES OF SOLUTIONS TO THE PHOSPHORUS SUSTAINABILITY CHALLENGE

The global future P scenarios presented in Figure 9.2 (based on Cordell et al. 2009) illustrate that meeting P demand by the year 2100 will require new efficiencies (i.e., demand-side measures) to reduce BAU P consumption by two-thirds. Recycling and reuse (i.e., supply-side measures) may have to make up the additional third of the future P deficit. In this section we present specific examples of both types of solutions (expanding from Cordell et al. 2009), with the caveat that we fully recognize that many solutions have likely not yet been thought up or invented.

DEMAND-SIDE MEASURES

Demand-side measures (or demand management) include any measures taken to reduce P waste through the food production and supply chain. These solutions will involve increases in P-use efficiency, primarily in agriculture, mining, food production, as well as reductions in our overall demand for P. We divide demand-side measures into two main areas: a) Solutions related to efficiency in agriculture and the food chain, and b) Solutions related to changes in human behavior and values that reduce individual and societal demand for P.

Agricultural Efficiency

To keep pace with a growing BAU population and a BAU dietary trajectory of higher global meat consumption, global grain production would have to double by 2050. To achieve agricultural sustainability, *crop and livestock production must increase without an increase in the negative environmental impacts associated with agriculture* (Tilman et al. 2002). To meet this increasing BAU demand, crop yields on existing agricultural land would have to increase in order to prevent the further conversion of wild areas to agricultural production. Current agro-industrial dogma suggests that this can only happen with increasing use of P fertilizers, which is in clear conflict with demand-side measures that seek to reduce P-use in agriculture. One option may be in "precision farming," where temporal and spatial nutrient supplies are precisely matched to crop demand. In precision farming, soils are intensively managed with frequent soil P testing, and fertilizer application is regularly adjusted in both location and timing. Satellite imagery is used to determine where fertilizer is needed in a field; fertilizers are applied directly to the root zone. Rolling out precision farming techniques to all scales of farming may lead to substantial reductions in P waste (and hence demand) in agriculture. Increasing the bioavailability of P in agricultural soils is also important. This may be accomplished with microbial inoculants that enhance P bioavailability and uptake in soils, allowing plants to more efficiently "mine" P from the soil reservoir. Most P lost from croplands is lost through soil erosion and storm runoff. Measures to reduce these losses, and to keep P in the soils and out of waterways, are often simple. Moving away from "P hungry" crops and animals to those that optimize P-use will likely involve genetic solutions, which may present other social and environmental challenges (see chapters 4 and 7 for details).

Solutions that involve the use of organic wastes (e.g., manures, digestate, human excreta) to replace phosphate rock-derived fertilizers cross both demand- and supply-side measures. Developing strategies that synchronize nutrient release from organic sources with plant demand will be essential if organic sources of P are to be truly effective in our current agro-industrial systems. A major barrier to reusing manure and digestate in crop production is the spatial separation of animal production and crop production. Manure is heavy and this severely limits the radius to which it can be transported to be reused; the co-location of crop and livestock production is the most viable way to optimize manure recycling to crop production, but as we note above this will require transformative reconceptualization of the current agro-industrial system (see Schroder et al. 2011 and chapter 5). The same argument can be made for P-use efficiency throughout the food chain. The co-location of livestock and crop production with food processing and consumption will further reduce P waste, thus reducing demand for "new" P as mineral-based fertilizers.

Changing Social Behavior and Societal Norms

More people need more food, but increasing socioeconomic status (i.e., affluence) may have an even greater per capita impact on P demand by increasing meat consumption. A key demand-side measure is the changing of diets from the meat-intensive lifestyles, currently seen as normal for affluent countries, to an increasing level of plant-based diets. Education is important; educational initiatives such as "one vegetarian day a week" will be needed worldwide. Moreover, reducing the amount of food consumed per person is important. Most individuals in affluent countries far exceed their required daily food (and thus P) intake. This introduces the concept of P footprinting, which will enable us to measure and compare demand for "new" P as a first step toward possibly conceptualizing a global P "quota" per person (see chapter 7).

SUPPLY-SIDE MEASURES

Supply-side measures are those that seek to meet demand with alternative sources of P and keep P cycling tightly within the terrestrial system, rather than being lost to water bodies. It also includes measures taken to increase the overall supply of "new" P from mining, where necessary (see chapter 3).

Reuse and Recovery of Phosphorus from Organic Wastes

All organic wastes contain P. Unless animals are actively growing, virtually all of the P that they consume is excreted and can be reused or recovered from manure or urine. Food wasted during consumption contains P, as do crop residues such as straw and stover. In the demand-side measures section we have already touched, there are geographical barriers to reusing manure on farmland. Manure contains a high percentage of water and is thus energy-intensive to transport. However, extracting P from these wastes, for example through extraction and precipitation as struvite, may enable P recovery in a crystalline, inorganic form that can be stored and transported more efficiently. Moreover, struvite also contains nitrogen, which brings up another important consideration.

All organic wastes contain resources other than P. Sustainable resource solutions that increase P supply must also facilitate recovery of other resources, such as nitrogen. Anaerobic digestion of food and agricultural wastes recovers energy as biogas and results in a digestate that can be separated into a nitrogen-rich liquid fraction and a P-rich fiber fraction, both of which may be reused in agriculture. Alternatively, struvite may be precipitated from digestate. To facilitate agricultural reuse of fresh manure or anaerobically digested manure, some agricultural areas have set up manure exchange markets, where farmers exchange organic waste or bid for them using an Internet-based auction system (see Cordell et al. 2011 for a spectrum of low- to high-tech recovery and reuse options throughout the food system). An advantage of increasing P supply from organic wastes is thus the ability to grow local P markets, encouraging the establishment of green industry and markedly increasing national P security. To encourage local reuse of P in organic wastes, P provenance may be highlighted and brought to the attention of the public in a "local food, local fertilizer" campaign (see chapter 7 for details).

Mineral Phosphorus Supply

Supply of phosphate rock that is economically extractable is finite and reserves in many areas are likely to run low this century. Moreover, if mining in China and the United States continues at current rates, these reserves are likely to be depleted by mid-century (or earlier for the United States), leaving global phosphate rock supply dependent largely on Morocco. Is it possible to increase the supply of phosphate rock? Exploration to identify more phosphate reserves is likely to continue, and technological advances may make it possible that more economically extractable reserves will be identified (e.g., off the coasts of Angola and Namibia). However, even this technological innovation is likely to require more energy expenditure to produce the same amount of usable P. If demand for P continues to increase—that is, if the sustainable solutions we propose here are not undertaken—prices will increase, which will have two contrasting effects: 1) this could add new reserves that are currently not economically viable for extraction, which will increase the supply of phosphate rock; 2) the higher prices will place P fertilizers further out of reach of many farmers in developing countries, exacerbating the geoeconomic inequities and geopolitical challenges that currently exist (chapter 3).

THE POLICY PALETTE: IMPLEMENTING SOLUTIONS TO THE PHOSPHORUS SUSTAINABILITY CHALLENGE

In this chapter we have presented a proof-of-concept for how to move forward with solutions to P sustainability challenges that is grounded in sustainability science theory and application. Given the global scale of the P challenge, and the fact that it requires environmental, social, and economic integration for solutions, policy making must be central to any "way forward." To this end, we present a palette to help guide policy-based implementation of solutions. This approach acknowledges barriers to transformative policy change, the identification and roles of key stakeholders, and possible instruments of viable sustainable P policy. There are numerous barriers to policy change, which we earlier referred to as institutional inertia, and we have attempted to summarize these into seven "deadly sins," per Dunstan et al. 2008 and Rutovitz and Dunstan 2009 (Figure 9.3; chapter 7). The central barrier to changing any policy is confusion. Confusion comes from ignorance, which is a product of a lack of available information or mis-education. Split incentives generate confusion through disarray. For example, when a sewage treatment plant operator benefits from reduced costs if P concentrations in effluent meet standards, these savings are not necessarily passed on to a homeowner who has installed a urine-diverting toilet. High discount rates cause economic impatience with the future and confusion about the present. Externalities and artificial pricing structures, often due to the lack

The Future of Global Phosphorus Sustainability

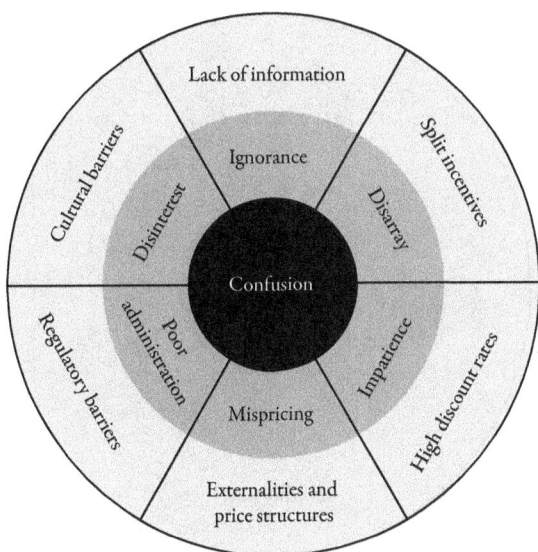

FIGURE 9.3 The seven "deadly sins" that represent barriers to changing policy and management of resources, including P (modified from Dunstan 2008, 2009). The outer light gray ring shows the policy barriers, or the "sins." The middle ring represents examples of how these policy barriers affect the current system, and the center ring is the result of these effects: system confusion.

of pricing of externalities, such as water pollution, are significant barriers to efficient resource use. Regulatory barriers put up by poor administration generate confusion, such as barriers to reuse of effluent or manures. Finally, cultural barriers to change result in disinterest, which quickly translates into confusion (e.g., a cultural resistance to changing diets to reduce consumption of meat and dairy products). None of these barriers are insurmountable—either alone or together—but recognizing that they must be confronted is clearly a first step to implementing policy changes for sustainable solutions.

We argue that the most viable direction forward on the sustainable management front should focus on a simple policy palette (Figure 9.4). In this palette, six policy strategies—information, facilitation, enticement, accurate pricing, useful regulation, and reasonable targets—are focused on a single policy goal: coordination. A critical aspect of the policy palette is that it sits on a template that includes environment, society, and economy. All of these critical sustainability concepts must be incorporated if policy change has any hope of transformational change. And the integrated and broad solutions that are necessary for successful solutions to the P sustainability challenge will almost certainly require more than a mere "tweaking" of the current systems, institutions, and infrastructures. We must not be afraid to envision transformational change that challenges the (sometimes comfortable) institutional and infrastructural inertias that are driving our systems, our lives, and our future on current trajectories.

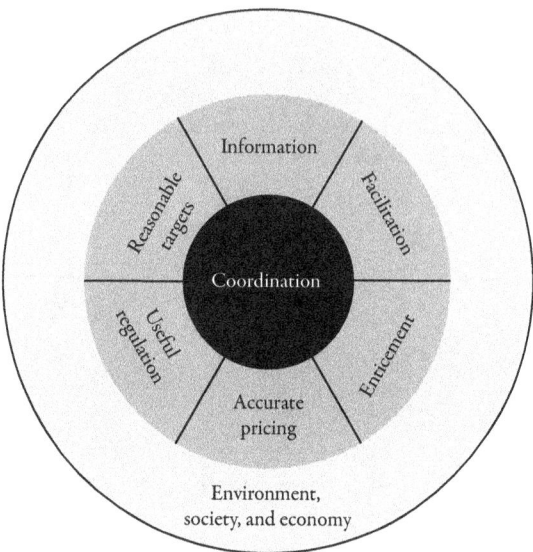

FIGURE 9.4 The policy palette based on the "seven deadly sins" of poor policy shown in Figure 9.3 (modified from Dunstan 2008, 2009). The outer light gray ring is the system itself. The middle ring represents examples of positive policy tools that correspond to the barriers in Figure 9.3, and the center ring is the result of these effects: cross-system coordination.

BOX 9.3
CHAPTER 9 SUMMARY

- This chapter identifies a vision for a sustainable future for phosphorus use, and documents a range of measures or actions that would move us toward that vision.
- We discuss several barriers to the implementation of these measures, and a range of policy tools that would help to overcome those barriers if they were to be applied in a coordinated way.

REFERENCES

Arnstein, S. R. 1969. A ladder of citizen participation. *Journal of the American Planning Association* 35:216–224.

Brewer, G. D. 2007. Inventing the future: scenarios, imagination, mastery and control. *Sustainability Science* 2:159–177. Springer Japan.

Childers, D. L., J. R. Corman, M. Edwards, and J. J. Elser, 2011. Sustainability challenges of phosphorus and food: Solutions from closing the human phosphorus cycle. *Bioscience* 61(2):117–124.

Clark, W. C., R. B. Mitchell, and D. W. Cash. 2006. Evaluating the Influence of Global Environmental Assessments. In *Global Environmental Assessments: Information and Influence*, edited by Ronald B. Mitchell, William C. Clark, David W. Cash, and Nancy M. Dickson, 1–28. MIT Press, Cambridge, MA.

Cordell, D. 2010. *The Story of Phosphorus: Sustainability implications of global phosphorus scarcity for food security*, PhD thesis. Linköping Studies in Arts & Sciences No. 509, Linköping University Press, Linköping, Sweden.

Cordell, D., A. Rosemarin, J. J. Schroder, and A. Smith. 2011. Towards global phosphorus security: A systems framework for phosphorus recovery and reuse options. *Chemosphere* 84:747–758.

Cordell, D., and S. White. 2009. *The Story of Phosphorus: Sustainability implications of global fertilizer scarcity for Australia—Synthesis paper*, National Workshop on the Future of Phosphorus, Sydney, 14th November 2008, Institute for Sustainable Futures, University of Technology, Sydney.

Cordell, D., S. White, J. O. Drangert, and T. S. S. Neset. 2009. *Preferred future phosphorus scenarios: A framework for meeting long-term phosphorus needs for global food demand*, International Conference on Nutrient Recovery from Wastewater Streams, Vancouver, Canada. Edited by Don Mavinic, Ken Ashley, and Fred Koch. IWA Publishing, London.

Costanza, R. 2000. Visions of alternative (unpredictable) futures and their use in policy analysis. *Conservation Ecology* 4:5–22.

Dreborg, K. (1996), Essence of Backcasting. *Futures* 28(9): 813–828.

Dunstan, C., K. R. Abeysuriya, and W. Shirley. 2008. *Win, win, win: Regulating electricity distribution networks for reliability, consumers and the environment: review of the NSW D-Factor and alternative mechanisms to encourage demand management*. Institute for Sustainable Futures, UTS, Sydney.

Fischer, F. 1993. Citizen participation and the democratization of policy expertise: From theoretical inquiry to practical cases. *Policy Sciences* 4:367–187.

Guston, D. H. 2008. Innovation policy: not just a jumbo shrimp. *Nature* 454:940–941.

Höjer, M. 2000. Determinism and backcasting in future studies. *Futures* 32:613–634.

Loorbach, D., and J. Rotmans. 2006. Managing transitions for sustainable development. In *Understanding Industrial Transformation: Views from Different Disciplines*, edited by X. Olshoorn and A. J., Wieczorek, 187–206. Springer, New York.

Midgley, G. 2003. Science as systemic intervention: Some implications of systems thinking and complexity for the philosophy of science. *Systemic Practice and Action Research* 16:77–97.

Olson, R. L. 1995. Sustainability as a social vision. *Journal of Social Issues* 51:15–35.

Raskin, P., T. Banuri, G. Gallopin, P. Gutman, A. Hammond, R. Kates, and R. Swart. 2002. *Great Transition: The Promise and Lure of the Times Ahead*. Stockholm Environment Institute, Boston.

Robinson, J. B. 1982. Energy backcasting A proposed method of policy analysis. *Energy Policy* 10:337–344. Elsevier.

Robinson, J., S. Burch, S. Talwar, M. O'Shea, and M. Walsh. 2011. Envisioning sustainability: Recent progress in the use of participatory backcasting approaches for sustainability research. *Technological Forecasting and Social Change* 78:756–768.

Robinson, J. B. 2003. Future subjunctive: backcasting as social learning. *Futures* 35:839–856.

Rutovitz, J., and C. Dunstan. 2009. *Meeting NSW electricity needs in a carbon constrained world: lowering costs and emissions with distributed energy*. Intelligent Grid Cluster, Sydney, Australia.

Schröder, J. J., A. L. Smit, D. Cordell, and A. Rosemarin, A. 2011. Improved phosphorus use efficiency in agriculture: a key requirement for its sustainable use. *Chemosphere* 84(6):822–831.

Swart, R., P. Raskin, and J. B. Robinson. 2004. The problem of the future: sustainability science and scenario analysis. *Global Environmental Change* 14:137–146.

Talwar, S., A. Wiek, and J. Robinson. 2011. User engagement in sustainability research. *Science and Public Policy* 38:379–390.

Tilman D., K. G. Cassman, P. A. Matson, R. Naylor, and S. Polasky 2002. Agricultural sustainability and intensive production practices. *Nature* 418: 671–677.

van Kerkhoff, L., and L. Lebel. 2006. Linking Knowledge and Action for Sustainable Development. *Annual Review of Environment and Resources* 31:445–477.

10

Concluding Remarks: Synthesis and Initial Steps toward a Sustainable Phosphorus Future

P IS FOR PLANNING

Jessica R. Corman, Karl A. Wyant, James J. Elser

Patricia Sahertian; Top: *enables him to cultivate; farmers' fertilizer company; a matter of highest importance;* Middle: *nelson's pop corn; "p" pee; efforts of science for its improvement;* Bottom: *desirous and peculiar; control over animal life;* Mixed media collage on paper, all are 5" x 3.5", 2010

Patricia Sahertian
Minding Your "P"s (eight-piece series)

Scientific Collaborators:
Elizabeth Cook, PhD student, School of Life Sciences, Arizona State University
Rebecca Hale, PhD student, School of Life Sciences, Arizona State University
David Iwaniec, PhD student, School of Sustainability, Arizona State University

Description of Artwork:
Artist Patricia Sahertian had no idea what she would be creating when, a little intimidated, she signed up to participate in the Sustainable P Summit's "Phosphorus, food, and our future: A collaboration between artists and scientists." She thought, "Now, that is something really intriguing." She was assigned to work with Elizabeth Cook, Rebecca Hale, and David Iwaniec, PhD students working on the budget of phosphorus in the city of Phoenix. Sahertian felt a sigh of relief when she realized she already knew Elizabeth through mutual friends. There was no need for the worried anticipation, for when she did get to meet with her scientists, she found they were helpful and enthusiastic about explaining their work and the role of phosphorus in terms that were easy to understand.

Cook, Hale, and Iwaniec are collaborators with other scientists on a project examining natural and human-mediated phosphorus flows and storage in the Phoenix metropolitan area. Fertilizer for agriculture and urban landscaping, as well as human foods, represents the largest flux of P to the city. Although export of P is high, reuse of wastewater recycles many nutrients back to the urban landscape and aids in preventing downstream environmental degradation.

Sahertian, having a storytelling nature, processed much of what she learned: that phosphorus is mined, that it is used in fertilizer, that we eat it and expel it, and finally that it can actually be recycled, and if not, it can have devastating effects on downstream ecosystems. She asked, "How can we show this process in a graphic way, without having to use too much text?" They discussed together abstract painterly ways of illustrating the levels of P, a flipbook idea, and even a flash animation. When Sahertian actually sat down to work on this project, though, some of these ideas seemed too vague or too complex to form the narrative. Working with collage as a medium, Patricia decided that it worked.

Doing some research into the history of phosphorus, Sahertian found old fertilizer brochures and articles about phosphorus printed in the *New York Times* that dated back to the 1800s. In addition, she incorporated some retro wallpaper and advertising images from the 1950s and '60s into the collages. One item, a vintage Premium King Korn Stamp, ties them all together. "Sifting through my collection of ephemera, I thought the King Korn stamp was very effective in conveying not only the image of agriculture, but also the exchange of money and the reciprocation of the gifts you get back when you 'cash' them in. Somehow that symbolized, for me, the whole recycling process," she said.

About the Artist:
Patricia Sahertian was raised right on the border of Brooklyn and Queens, the daughter of Irish and Greek immigrants. Influenced by her visits to the Metropolitan Museum of Art, the New York Public Library, and the 1964 World's Fair, she developed a love of visual art at an early age. She has studied painting, printmaking, ceramics, and sculpture.

Her fascination with found objects, old letters, and late-nineteenth-century history is evident in her current work, as is her extensive experience in graphic and multimedia design and film.

Sahertian considers herself a storyteller in picture form and is focused on developing her narrative style. She also reads coffee grounds.

BOX 10.1
CHAPTER 10 OBJECTIVES

In this chapter we seek to integrate the perspectives presented in this book, leading to an overall view:

- to integrate and synthesize the main themes and findings from the book in order to help place the P sustainability challenge in the broader context of sustainability science;
- to identify missing pieces and gaps in these analyses and in the overall P sustainability efforts articulated here and elsewhere;
- to identify some promising areas for future progress in P sustainability science as well as for individual action to enhance P sustainability.

PHOSPHORUS SUSTAINABILITY: A CASE STUDY FOR SUSTAINABILITY SCIENCE

The contents of this book show that achieving phosphorus sustainability is, indeed, a "wicked problem" (*sensu* Conklin 2006) and well-suited for sustainability science. Conversely, the diversity of challenges presented, and their complex interfaces, suggests that the study of phosphorus sustainability may inform approaches used in sustainability science as a whole. As discussed in chapter 1, *sustainability* is defined as "the need to ensure a better quality of life for all, now and into the future, in a just and equitable manner" (Agyeman et al. 2003). Phosphorus is a resource essential for life (chapter 2), and each chapter in this book illustrates various dimensions by which current phosphorus-use does not meet Agyeman's definition. Going further, different chapters touch upon Gibson's (2006) criteria that define issues best suited for a sustainability science perspective (Table 10.1): socio-ecological integrity is discussed in chapters 2, 5, and 7, adaptation is discussed in chapter 6, and long-term planning and integration is discussed in chapter 9, among others.

In chapter 1, the authors present a visualization of the keystones of sustainability thinking (environment, society, and economy) and populate the diagram with concepts integral to the phosphorus sustainability challenge (e.g., "eutrophication" in Environment, "dietary preferences" in Society, and "price of P" in Economy). While writing their chapters, we asked each set of authors to identify how their chapter's concepts mapped onto sustainability thinking, using this socio-ecological framework (Figure 1.2). We compile these choices below (Table 10.1).

While some chapters focus on only one sphere of the problem, it is easy to find links to the other spheres within these chapters. For instance, chapter 2 covers issues and concepts mainly within the Environment space, but its discussion of eutrophication extends to include wastewater treatment (Economy) and the value of water for recreation and other cultural activities (Society). In chapter 4, the authors present different technologies that may improve phosphorus use in agricultural systems (Society) but also mention how ambient soil conditions can influence these technologies (Environment). And chapter 7 focuses mostly on the Society space, but it includes a discussion of how societal choice influences the biogeochemical cycling of phosphorus (Environment). Indeed, many of the chapters' domains cross multiple spheres, although this overlap is not always explicitly made. Minding such connections is imperative to phosphorus sustainability, as well as any sustainability problem.

Let's consider one of the most obvious and well-known consequences of wasteful phosphorus use: eutrophication. The authors of chapter 1 define "eutrophication" within the Environment space of the diagram, as inherent physical,

TABLE 10.1

TOPICS FROM SPHERE COVERED WITHIN EACH CHAPTER

	Environment	Society	Economy
2	Essential mineral, biogeochemical cycling, eutrophication		
3	Limited P minerals		Geopolitical tensions, price of P
4		Agricultural production	Infrastructure
5			
6	Limited P mineral, eutrophication	Future, waste	
7		Cultural values, waste management, dietary preference, agricultural production, food security	
8			Complex interactions, agricultural production, education
9	All	All	All

geological, and hydrological characteristics of a lake and its environs that can influence the lake's resilience to anthropogenic activities such as nutrient pollution. However, characteristics related to the environment are not the whole story for why some lakes turn eutrophic in response to increased anthropogenic activities while others do not. We must also consider eutrophication within the greater socio-ecological system; politics and management are influential. As discussed in chapter 5, adaptable wastewater treatment infrastructure in the watershed of a lake (e.g., Lake Nahual Huapi, Argentina) can greatly mitigate pressures from population increases that might otherwise lead to increased nutrient loading and water-quality degradation.

In considering the book as a whole, you can see that there is a shift from focusing on the Environment space earlier in the book to the Society and Economy spaces in the later chapters. In noting this we do not mean to suggest that there should be a linear approach that proscribes one to think only about the environment when discussing phosphorus cycling and only about the economy when thinking about future scenarios of phosphorus use (indeed, as reflected in chapters 8 and 9, a comprehensive approach to future planning is needed). The Environment, Society, and Economy categories are useful for framing discussions, but they should not be considered borders or boxes from which discussions cannot stray.

Minding the connections within the circle of sustainability also strengthens the link between the potential strategies for phosphorus sustainability and the social context within which those strategies may be implemented. For instance, the use of genetically modified organisms (GMOs) with increased phosphorus use efficiencies may be an appropriate strategy to increase efficient use of fertilizers and decrease phosphorus runoff in agriculture (chapter 4). However, religious, political, and/or cultural norms may impede their implementation (chapter 7). Likewise, "end-of-pipe" technologies for more efficient phosphorus recycling in wastewater streams face, at times, significant cultural and financial barriers (chapters 6 and 7).

Interestingly, while the authors of chapter 1 list human population size as a phosphorus sustainability issue, it is not the explicit focus of any chapter (although it is mentioned in many chapters: 3, 5, 6, 8, and 9). Food security for a growing population has been a driver behind worries of phosphorus supplies (Cordell et al. 2009) and such concerns will continue to grow with increasing population pressures. To achieve phosphorus sustainability, we will have to address human population size more explicitly. This is not a problem unique to phosphorus sustainability; indeed, it is central to most other socio-ecological challenges. And, as with these other challenges, the very same places where population sizes are increasing at the greatest rates are the places where involvement of the government or other representatives is under-represented in the discourse (see below).

CHANGING THE DIALOGUE

The bulk of this book reflects an attempt to move the discussion beyond the current paradigm in which phosphorus is on a "one-way trip"—from mine to food to fork—to one in which there is a sustainable, largely closed "human phosphorus loop" (Childers et al. 2011). While we hope that this book will help shift awareness and practice of phosphorus sustainability, we recognize that there are still important gaps in the discourse.

Recent (and not-so-recent—see Box 10.2) conversations and press about phosphorus sustainability have centered around one point: the possibility that high-quality, easily mined, and inexpensive phosphate rock is quickly becoming scarce. This book purposely was *not* structured around this point, but instead attempted

> BOX 10.2
>
> In 1938, U.S. President Franklin D. Roosevelt addressed the phosphorus sustainability issue in a statement to Congress: "The phosphorus content of our land, following generations of cultivation, has greatly diminished. It needs replenishing. I cannot overemphasize the importance of phosphorus not only to agriculture and soil conservation but also to the physical health and economic security of the people of the nation. Many of our soil deposits are deficient in phosphorous, thus causing low yield and poor quality of crops and pastures."

to embed this issue within the broader set of circumstances and conditions that make phosphorus use a "wicked problem." Therefore, there is only one chapter that tackles the question of phosphate rock and reserves per se (chapter 3). Within this chapter, we learn that quantity itself is not the primary problem; instead we see that neither phosphate rock nor phosphate fertilizer is distributed homogeneously around the globe, leading into the issues of geographical accessibility and affordability (embedded within the "society" sphere and the "economy" sphere, respectively).

There are still pieces of the dialogue that are missing. A central challenge of sustainable phosphorus is that the discourse must engage an enormous variety of participants (e.g., operations and mine owners, farmers and fertilizer producers, food processors and retailers, water quality managers, and, of course, anyone who eats and excretes) and must encompass the diverse local, regional, and global contexts in which the interactions between these participants take place. One of the goals of this book was not only to identify the actors involved throughout the phosphorus supply chain: mines (chapter 3); agricultural operations (chapters 4 and 5); waste processing systems (chapter 6), but also to describe these actors within the greater network of shifting economic and social demands on agricultural products (chapter 7).

We see in this book that the geographically disparate and often indirectly linked actors in the phosphorus supply chain often operate in isolation with little bearing on each other's activities. Yet, decisions regarding phosphorus use in one domain can often have profound effects on the activities and capacity of another domain.

Presently there are several global efforts to bring these different actors together, including a recurring international conference devoted to sustainable phosphorus thinking (the "Sustainable Phosphorus Summit": Linkshoping, Sweden (2010); Tempe, Arizona (2011); Sydney, NSW, Australia (2012)). These meetings have been well-attended by many at various steps of the phosphorus supply-and-demand system and have often included speakers from the higher echelons of the mining, fertilizer, and wastewater treatment industries. Rounding out the attendance has been a variety of academics (natural resource economists, ecologists, agronomists, etc.) and representatives from the nonprofit sectors. Indeed, conference representatives account for each step of phosphorus's journey from the mine to the drain (as well as possible steps from the drain back to the farm). However, it fair to say that more attention needs to be paid to the developing world.

As discussed in chapters 4 and 5, abundant food sources from the "Green Revolution" are to a large extent made possible thanks to phosphorus fertilizers and their augmentative effect on plant growth and production (World Bank/FAO 1996). In fact, over the last 30 years, a period during which world population nearly doubled, increased fertilizer use has helped raise average per capita consumption of food by 17 percent to 2760 kcal per day (Pretty et al. 2003). Despite the productivity gains afforded by fertilizer use, many countries still face persistent challenges in food security. There are an estimated 800 million people lacking adequate access to food, particularly in Africa and southeast Asia (Uphoff 2002). Indeed, roughly 33 countries still have average per capita food consumption of less than 2200 kcal per day, below recommended intake levels (FAO 2000). Many of the regions facing food security are the same ones that are not at the table during phosphorus sustainability discussions. When the limited geographic diversity is coupled with the lack of attention to increasing population size (see previous section), there is a very real challenge to current efforts toward phosphorus sustainability.

In light of the enduring problem of global hunger and low agricultural productivity in various parts of the developing world, we think it's important to emphasize here that these regions of the world actually require *more* P inputs into their soil and agricultural systems so that they may alleviate food shortages and ensure a healthy and productive citizenry (chapter 5). While much of the material discussed in this book calls attention to the need for *lowering* the overall intensity of human phosphorus use, such recommendations should be seen as region-specific, as should discussions of various means for phosphorus recovery and recycling. For example, discussions in more developed economies often include reducing phosphorus inputs into soils or recovering phosphorus, in the form of struvite, from large-scale wastewater treatment plants. Such "solutions" to phosphorus sustainability ignore the plight of regions that do not benefit from a historical legacy of fertilizer over-application and storage in soils or do not have the working capital and/or infrastructure to support centralized P recovery systems. Details of some "low-cost, low-tech" phosphorus recovery strategies are described in chapter 7 and should become a standard talking point in the discussion of phosphorus sustainability.

INSTITUTIONAL GAPS

One particularly challenging aspect of phosphorus sustainability is the institutional gaps that relate to governance of phosphorus extraction, application, and impact (Cordell et al. 2009). Also notable is the general lack of any institutional policies or structures charged explicitly with assessing and assuring the long-term supply, affordability, and distribution of phosphorus at the global scale (Cordell et al. 2009). In this regard the institutional challenges of phosphorus sustainability per se are little different from many of those facing society across all domains. While these issues were touched upon in different chapters (e.g., fertilizer markets are global in scale, but the impacts of fertilizer runoff on ecosystems are felt regionally and locally (chapters 2 and 5)), we will take a few paragraphs to delve more deeply into the issue.

Institutional gaps in governance for sustainability were highlighted recently by Biermann et al. (2012). The authors stressed a variety of changes in large-scale institutions, such as the United Nations and the World Bank, that are needed to address the pressing challenges of global sustainability. Some of these have relevance to phosphorus sustainability in particular, including the importance of more rigorously integrating environmental dimensions into sustainable development efforts (e.g., developing world farmers need affordable phosphorus fertilizers to increase their yields and thus their affluence, but they also need clean, abundant drinking water) and the potential for global trade mechanisms to enhance environmental sustainability (e.g., trade agreements involving food could incorporate incentives to favor use of recycled over mined phosphorus). However, the ineffectiveness of inter-governmental institutions and mechanisms to mitigate the pressing, high-profile sustainability challenge of global climate change discourages optimism that such high-level approaches will be effective for achieving phosphorus sustainability.

Beyond such global governance issues, Biermann et al. (2012) also call attention to the potential importance of human behavioral changes, small-scale civic organizations, and the private sector in addressing sustainability, especially if larger-scale structures are in place to allow such changes to be effective. In the case of phosphorus, and in contrast to the case of climate change, we think that there is reason for some optimism. At least some of this optimism reflects biogeochemical differences: carbon circulates globally in the relatively well-mixed atmosphere, where it has a long residence time and affects a slow biological-time-scale variable, global average temperature. Thus, the benefits of any small-scale effort made by individuals, communities, or regional governments can be hard to perceive and diluted into the large reservoir of others' inactions in affecting atmospheric chemical composition. In contrast, phosphorus cycling lacks such large-scale circulation in a well-mixed reservoir (chapter 2). Instead, phosphorus tends to cycle locally and regionally at scales closer to those of existing institutions for human governance (watersheds). Phosphorus use also has impacts (green, unpleasant lakes; dead fish; chapters 2 and 5) that can be quite clearly perceived if knowledge of the phosphorus-eutrophication link is widespread. Importantly, phosphorus also tends to have shorter ecosystem residence times; thus, the benefits from reduced phosphorus runoff can be felt locally and within timescales relevant to human perceptions and attitudes, including those who made the (perceived) sacrifices.

MULTIPLE BENEFITS OF PHOSPHORUS SUSTAINABILITY: FOOD FOR THOUGHT

Intriguingly, some efforts that may be needed to increase global phosphorus sustainability map directly to those that may also enhance individual welfare. Consider diet (chapter 7). Reducing global phosphorus demand to reduce eutrophication and to enhance phosphorus affordability will likely require reducing the individual "P footprint" of the diet, which strongly reflects the prevalence of meat, especially beef (Cordell et al. 2009). Thus, achieving global phosphorus sustainability may require cultural shifts in which people eat lower on the food chain (chapter 7). Fortunately, motivations for such a shift need not necessarily come only from environmental altruism (or even concerns about animal welfare)—low-meat diets (and especially those with little red meat) are significantly healthier in terms of effects on heart disease, cancer, and other chronic illnesses (Zheng et al. 2009, Sinha et al. 2009, Pan et al. 2012). Indeed, there is provocative evidence that high-P diets per se (independent of meat content) complicate kidney dialysis treatment and may amplify aging, cardiovascular disease, and some forms of cancer (Elser et al. in press). This alignment of individual and collective interest, as well as the closer correspondence between scales of existing governance and of phosphorus cycling mentioned above, suggests that "bottom-up" measures to enhance phosphorus sustainability may indeed be successful, regardless of the effectiveness of global-scale institutional structures. We also wonder if the emerging predominance of social media may empower citizens and local civic structures to be particularly effective in implementing measures to move toward phosphorus sustainability, via both market forces and local and regional governmental actions.

WHERE DO WE GO FROM HERE?

When it comes to phosphorus sustainability, sustainability science points at the obvious: there is no "silver bullet." While strategies discussed in this book, such as shifts in diets (see section above; chapter 7) or changes to wastewater technology (chapter 6), may be essential in different regions, none can be considered a panacea. As discussed in chapters 1 and 9, sustainability science offers a transdisciplinary framework for approaching complex socio-ecological problems by taking advantage of multiple stakeholder engagement. It approaches "problem-solving" by taking a nonlinear and iterative approach, incorporating changing stakeholder views and reevaluating priorities through time. A generalized road map for phosphorus sustainability is proposed in chapter 9: "backcasting." Backcasting determines a path forward to a future sustainable state by beginning at that desired state and moving backwards in time. It incorporates stakeholder views and helps identify intervention points. It is also unimpaired by the current state of the system. However, as discussed in chapter 9, backcasting is still influenced by the current system: while it provides a useful framework for discussions of change, stakeholders are still left with the challenging task of overcoming the inertia of the current system. While this approach is not easy (a moving target is always more difficult!), we believe it is imperative for creating a sustainable phosphorus future.

Of course, despite these complexities, there are still things you can do as an individual to change our current unsustainable phosphorus system. As we discussed above, the phosphorus sustainability challenge has several features in which individual and small-scale benefits may also map to larger-scale outcomes. Here are some ideas:

- Work to reduce your individual P footprint by eating lower on the food chain and replacing P-intensive proteins (e.g., beef) with low-P proteins (e.g., fish, chicken, or beans). As discussed above, doing so will probably be good for your waistline and your cardiovascular system, too.
- Support local efforts for phosphorus recycling in municipal wastewater treatment if such efforts are under way.
- Reduce your household phosphorus use by avoiding phosphate-based detergents and lowering or eliminating fertilizer application to lawns and landscaping.
- Discuss phosphorus sustainability issues with your local food suppliers, including grocers, farmers' market purveyors, and restaurant owners.
- Grow your own food using composted fertilizers.

- Consider installation of emerging phosphorus recycling sanitary systems in your home and encourage their adoption in your neighborhood or town (see the Sustainable Sanitation Alliance for information and ideas: http://www.susana.org);
- Join the global discussion of phosphorus sustainability through the Global Phosphorus Network (GPN; http://globalpnetwork.net), the ASU Sustainable Phosphorus Initiative (http://sustainablep.asu.edu), the Global Phosphorus Research Institute (GPRI; http://phosphorusfutures.net), the Dutch Nutrient Platform (http://www.nutrientplatform.org/), and GlobalTRaPS (www.uns.ethz.ch/gt), among others.

BOX 10.3
CHAPTER 10 SUMMARY

- Global use of phosphorus represents a "wicked" problem.
- Phosphorus sustainability is a hallmark sustainability challenge.
- Better inclusion of developing nations and emerging economies is needed in the discussion of phosphorus sustainability challenges and potential strategies.
- There is no "silver bullet" for phosphorus sustainability, but there exist many different strategies that may be appropriate in different contexts.

REFERENCES

Agyeman, J., R. D. Bullard, and B. Evans. 2003. *Just sustainabilities: development in an unequal world*. The MIT Press, London.

Biermann, F., K. Abbott, S. Andresen, K. Baackstrand, S. Bernstein, M. M. Betsill, H. Bulkeley, B. Cashore, J. Clapp, C. Folke, A. Gupta, J. Gupta, P. M. Haas, A. Jordan, N. Kanie, T. Kluvankova-Oravska, L. Lebel, D. Liverman, J. Meadowcroft, R. B. Mitchell, P. Newell, S. Oberthür, L. Olsson, P. Pattberg, R. Sanchez-Rodriguez, H. Schroeder, A. Underdal, S. C. Vieira, C. Vogel, O. R. Young, A. Brock, and R. Zondervan. 2012. Navigating the Anthropocene: Improving Earth system governance. *Science* 335:1306–1307.

Childers, D. L., J. Corman, M. Edwards, J. J. Elser. 2011. Sustainability Challenges of Phosphorus and Food: Solutions from Closing the Human Phosphorus Cycle. *BioScience* 61 (2):117–123.

Conklin, E. J. 2006. Dialogue mapping: building shared understanding of wicked problems. Wiley, Chichester, England; Hoboken, NJ.

Cordell, D., J.-O. Drangert, and S. White. 2009. The story of phosphorus: Global food security and food for thought. *Global Environmental Change* 19:292–305.

Elser, J. J. in press. Health dimensions of phosphorus. In *Sustainable Phosphorus Management: a Transdisciplinary Roadmap*, edited by R. W. Scholz, A. H. Roy, F. S. Brand, D. T. Hellums, and A. E. Ulrich. Springer Book.

FAO, 2000. *Agriculture: towards 2015/30*. Global Perspectives Studies Unit. FAO, Rome.

Pan A., Q. Sun, A. M. Bernstein, M. B. Schulze, J. E. Manson, M. J. Stampfer, W. C. Willet, and F. B. Hu. 2012. Red meat consumption and mortality: results from 2 prospective cohort studies. *Archives of Internal Medicine* 172(7):555–563.

Pretty, J. N., J. I. L. Morson, and R. E. Hine. 2003. Reducing food poverty by increasing agricultural sustainability in developing countries. *Agricultural Ecosystems and Environment* 95:217–234.

Sinha R., A. J. Cross, B. I. Graubard, M. F. Leitzmann, and A. Schatzkin. 2009. Meat intake and mortality: a prospective study of over half a million people. *Archives of Internal Medicine* 169(6):562–571.

Uphoff, N. 2002. *Agroecological Innovations: Increasing Food production with Participatory Development*. Earthscan, London.

World Bank/FAO. 1996. *Recapitalization of soil productivity in Sub-Saharan Africa*. World Bank/FAO. Washington DC/Rome.

Zheng W., and S. A. Lee. 2009. Well-done meat intake, heterocyclic amine exposure, and cancer risk. *Nutrition and Cancer* 61(4):437–446.

GLOSSARY

CHAPTER 1

CLOSED (LOOP OR SYSTEM): refers to a system where P is highly recycled or reused and there are minimal loses from the system.

EXTERNALITIES: consequences that are not included in the price of an item or service.

GREEN REVOLUTION: rapid increase of crop yields and thus food production in the 1970s due to important advances in research and technology. Most notable are the development and widespread use of high-yielding crop varieties coupled to expanded irrigation, fertilizer, and pesticide use.

HUMAN P cycle: Refers to how humans use P by mining P rocks, producing fertilizer, producing food, consuming it, and producing and handling waste.

OPEN (CYCLE OR SYSTEM): refers to a system that uses P in a "linear manner" where there are a lot of losses (or waste) and little recycling. This is how we currently use P resources.

P ACCESSIBILITY: the ability of an interested user to procure and apply forms of P to optimize food or industrial production given social, economic, and environmental constraints.

PEDOGENESIS: process of soil development.

SOCIAL-ECOLOGICAL SYSTEM: a system that is neither completely natural nor completely human-made. In such systems there are interactions between humans and the ecosystem, which means that neither component can be ignored in management.

SUSTAINABILITY: "the need to ensure a better quality of life for all, now and into the future, in a just and equitable manner, whilst living within the limits of supporting ecosystems" (Agyeman et al. 2003).

TRANSDISCIPLINARITY: A holistic approach to research that includes not only representatives from multiple disciplines of academia, but also policy makers and other practitioners that are involved in particular domain.

WICKED PROBLEM: a problem that is complex and urgent, exhibits long-term dynamics, involves cross-sectoral and cross-scalar interactions, and often has solutions that are place-based.

CHAPTER 2

APATITE: The mineral form of phosphate found in bones and teeth, as $Ca_{10}(PO_4)_6(OH)_2$; similar mineral P is also found in rocks and soils.

DNA AND RNA: Deoxy-ribonucleic acid and ribonucleic acid; biological polymers made up of repeating units (nucleotides) that each contains a phosphate moiety. DNA is the basis of genetic inheritance, while various forms of RNA are involved in regulation and production of proteins needed for cellular growth and function.

ELEMENT: One of the more than 100 substances (of which P is one) that cannot be broken down or changed into another substance using chemical means; each element represents a particular stable configuration of sub-atomic particles (protons, neutrons, electrons).

EUTROPHICATION: The process of increasing ecosystem productivity and biomass growth over time; in lakes and oceans, increased inputs of limiting nutrients (N, P) from human activities (e.g., fertilizer runoff, sewage inputs) accelerate this process and produce blooms of undesirable algae and depletion of oxygen in lower waters (hypoxic zones), resulting in mortality of fish and invertebrates.

HYPOXIC ZONE: Lake or ocean water layers in which oxygen concentrations are depleted; the size of hypoxic zones and the degree of oxygen depletion are increased due to anthropogenic nutrient inputs that stimulate algal production in overlying waters. When the algae die and sediment to lower layers, their remains are decomposed by microbes, consuming oxygen.

IMMOBILIZATION: Microbial production of organic P compounds that are less available to plants.

MINERALIZATION: Microbe-mediated release of PO_4-P from organically bound forms into soil solution.

OCCLUSION: The physical/chemical processes by which P becomes unavailable in soils by chemical precipitation into various minerals involving Ca, Mg, Al, and/or Fe; occlusion can also involve physical protection from dissolution or plant uptake.

PHOSPHATE: PO_4^{3-}; the most common molecular form in which phosphorus is found in nature; in this form; a P atom is chemically bound to four oxygen atoms.

PHOSPHORUS TRANSFER CONTINUUM: The coupled steps by which phosphorus is supplied (from various sources, such as fertilizer input or primary rocks), is made available for transfer ("mobilized," via solubilization or detachment), is delivered (via surface or subsurface hydrologic transport), and ultimately affects receiving waters (streams, lakes, oceans).

PHYTATE: A chemical form of P formed in seeds and grains; this form of P is not bioavailable to non-ruminant livestock (e.g., pigs) or poultry.

POLYPHOSPHATE: A long polymer of linked phosphate molecules; polyphosphates are formed by microbes as a form of P storage and can be important in P removal in wastewater treatment.

Glossary

CHAPTER 3

APATITE: The ore mineral of phosphorus, a natural calcium phosphate. As a commodity, rock phosphate.

COST STRUCTURE: The unique set of costs associated with producing a given final product; these costs are attributed individual factor inputs such as capital, labor, natural resources, or any sort of intermediate input necessary throughout the supply chain.

DIAGENESIS: Complex natural processes affecting a sediment after its initial deposition, as it is hardened into a rock. For phosphorite, diagenesis commonly results in enrichment in phosphorus.

HUBBERT CURVE: A mathematical model of production of a mineral commodity over time, where production depends on the fraction remaining to be produced; generally predicts a peak (e.g., "peak phosphorus") followed by a gradual (and symmetric) decline. Devised by M. King Hubbert with regard to petroleum production.

INTERMEDIATE PRODUCT: A product used in the production of another product; for example, steel is an intermediate product in the production of automobiles.

ORE: Rock containing a sufficient concentration of a mineral commodity (expressed as grade) to be mined at a profit.

PHOSPHATE ROCK: Rock containing enough apatite to be potentially mined at a profit (i.e., to constitute ore). The raw material for phosphate fertilizer production.

PHOSPHORITE: Sedimentary phosphate rock; occurs in layers or beds.

R/C RATIO: Ratio of reserves (amount) to the present consumption rate (amount/year), expressed in years. Also known as static lifetime.

RECOVERABLE CONCENTRATE: Amount of rock phosphate (i.e., mineral apatite) that can be recovered (produced) as a result of mining and beneficiation (artificial concentration) of a given phosphate rock ore reserve.

RESERVE (OR **ore reserve**): Amount of a resource that can be mined at a profit (i.e., that constitutes ore) at any given time.

RESERVE BASE (or resource base): Total amount of a resource that might conceivably become a reserve in the future, as conditions change.

RESOURCE PYRAMID (or triangle): Conceptual diagram showing the geological amount of a given mineral resource as the horizontal area or width and profitability (relative economic viability) as the height.

SUPPLY CHAIN: The network of direct and indirect steps necessary to produce a given final product.

CHAPTER 4

ASSIMILATION: The incorporation of carbon, nitrogen, phosphorus, and other elemental nutrients into cellular biomolecules (e.g., nucleic acids, amino acids, and structural compounds.

AUXIN: A plant hormone that regulates various functions, including cell division and elongation.

BAND APPLICATION: An application method in which farm materials such as fertilizer or herbicides are applied in strips, usually to a bed or to a seed row.

BEST MANAGEMENT PRACTICES (BMPs): Best Management Practices are methods that are designed to prevent or reduce the transport of pollutants from the land into receiving waters, and that are intended to protect both surface and groundwater quality from potential adverse effects of agriculture.

BROADCAST APPLICATION: The application of farm materials such as fertilizer or herbicides to the entire surface of a field.

CONSERVATION RESERVE PROGRAM (CRP): The Conservation Reserve Program (CRP) is a voluntary program for agricultural landowners in the United States that is funded by the U.S. Department of Agriculture. Through CRP, landowners can receive annual rental payments and cost-share assistance to establish long-term, resource-conserving plant covers on eligible farmland.

DRY DISTILLER'S GRAIN (DDG): Dry distiller's grain (DDG) is an edible, solid by-product of ethanol production from cereal grains such as corn or wheat.

FERTILIZER PHOSPHORUS-USE EFFICIENCY (FPUE): The mass of plant tissue, in grams of dry weight, that can be produced per unit fertilizer P applied.

H+-PPase A membrane-bound enzyme that translocates protons (H+) by utilizing the energy of a pyrophosphate (PPi) molecule.

PHOSPHATE ROCK DECISION SUPPORT SYSTEM (PRDSS): The PRDSS is a model that allows users to calculate the efficiency of phosphate rock use under a wide range of conditions.

PHOSPHORUS-USE EFFICIENCY (PUE): The mass of tissue, in grams of dry weight, that can be produced by a plant or animal per gram of assimilated phosphorus.

PHOTOSYNTHATE: Photosynthetically produced organic carbon.

PHYTASE: An enzyme that breaks down the phosphorus-rich organic compound phytate.

RHIZOSPHERE: The region of soil that is subject to the influence of plant roots.

RUMINANT: An animal that chews the cud regurgitated from its rumen, such as cattle.

STOVER: The dried stalks and leaves of a field crop (especially corn) that are used as animal fodder after the grain has been harvested.

VARIABLE RATE FERTILIZER APPLICATION (VRA): Adjustment of the amount of an input such as seed, fertilizer, lime, or pesticides in order to best meet local conditions and to maximize crop yield potential in a field.

CHAPTER 5

ACTIVATED SLUDGE PROCESS: A wastewater treatment in which oxygen and organisms (generally species of bacteria and protozoans) are added to wastewater. The microbes use the sludge and available oxygen to respire, decreasing the concentrations of organic matter, N, and P.

ALKALINE SOILS: Soils with a pH > 7. These soils are often found in arid ecosystems, where, with precipitation less than evapotranspiration, cations accumulate.

BIOSOLID: Residual solids, or sludge, following wastewater treatment that have been treated such that they may be reused as fertilizer.

CALCAREOUS SOILS: Soils containing relatively high concentrations of calcium carbonate ($CaCO_3$). These soils are generally alkaline and are common to arid ecosystems.

EUTROPHICATION: Stimulation of aquatic organisms through the addition of nutrients. Increased biological activity then reduces dissolved oxygen concentrations.

LIME: Substance used to reduce soil acidity (i.e., increase pH). Contains calcium or magnesium oxides and hydroxides.

MONOGASTRIC: Containing a single stomach chamber.

PHOSPHORUS (P) BALANCE: The difference between phosphorus inputs (usually inorganic fertilizer inputs) and phosphorus removed in harvested crops, livestock, or lost to leaching and erosion (P balance—P inputs—P outputs).

BIOAVAILABILITY: The ability of P to be accessed and assimilated by organisms.

SORPTION CAPACITY: The ability of soils to bind added phosphorus such that it remains bound by physical or chemical means in the soil column.

PHYTATE: The molecule that stores the most phosphorus in plant tissues [$C_6H_{18}O_{24}P_6$], including most plant tissue (e.g., seeds) used as livestock feed.

POLYMICTIC: Without thermal stratification. Polymictic lakes are shallow lakes that mix thoroughly from top to bottom, maintaining homogenous temperature throughout the water column for at least part of the year.

POPULATION EQUIVALENT: The amount of organic matter that is degraded biologically with an oxygen demand over five days (BOD5) equal to 60 grams per day (Norwegian Standard NS 9426).

RIPARIAN BUFFERS: Vegetation planted along stream banks to improve water quality by reducing erosion and increasing runoff interception, thus reducing nutrient and sediment loading to surface water.

SUBSOILING: The deep tilling of soil (typically about 30–40 cm below the soil surface). Subsoiling aerates soils as well as bringing nutrients and organic matter found below the topsoil into surficial soil layers.

CHAPTER 6

ANAEROBIC DIGESTION: Biological degradation process under oxygen depleted condition. Methane-rich biogas is produced as a by-product. It is widely used for stabilization of organic wastes and renewable energy production.

ADVANCED WASTEWATER TREATMENT PROCESS: Also known as tertiary process or "effluent polishing." It is to improve effluent quality to meet increasingly stringent discharge standard beyond conventional wastewater treatment process. It typically involves filtration, coagulation with chemical addition, and activated carbon adsorption, but also enhanced biological phosphorus removal process is regarded an advanced wastewater treatment process.

COMMUNICATIVE PLANNING: Planning practice based on information sharing among various stakeholders and consensus and dialog-based decision-making process. In such a process, planners are encouraged to take the initiative for joint fact-finding, setting a scope and goal, conflict resolution, and fostering collaboration.

CONSTRUCTED WETLANDS: Artificial wetland system designed to remove pollutants (e.g., solids, organics, nutrients). Pollutants are removed by sedimentation, absorption by soil, or plant uptake as influent water moves through the system with sufficient hydraulic retention time.

CONVENTIONAL WASTEWATER TREATMENT PROCESS: Wastewater treatment process utilizing preliminary treatment which provides screening and separation of large objects; grit and grease; primary treatment where organic matter is physically removed through settling (primary sludge); and secondary treatment where the remaining BOD is consumed. Microbes can grow in suspension (activated sludge or oxidation ditches) or as a fixed film (trickling filters or biofilm reactors).

ECOLOGICAL SANITATION (EcoSan): Alternative sanitation practice intended to mimic natural nutrient cycle, while providing affordable and culturally acceptable sanitation for all. Source separation (e.g., urine diversion) and treatment is one of the fundamentals.

GREEN BUILDING: Houses and buildings which are resource and energy efficient by design throughout their useful life cycle. There are several rating system to measure the environmental performance of buildings; one is Leadership in Energy and Environmental Design (LEED) by US Green Building Council and Living Building Challenge.

HUMAN EXCRETA: Human excreta are a mixture of urine and feces. Contact with human excreta is a major transmission pathway for infectious disease.

IMPROVED SANITATION: Sanitation practice recognized by WHO and UNICEF to provide an agreeable level of protection of public health, which includes connection to a public sewer, septic system, pour-flush latrine, dry toilet, and some form of pit latrine.

INDUSTRIAL SYMBIOSIS: System approach to improve the overall economical and environmental performance of production practice by sharing services, utilities, and by-product resources among several industrial sectors.

SEPTIC SYSTEM: Small-scale on-site wastewater treatment system. A septic system often consists of a septic tank where solids are settled out and stabilized anaerobically or aerobically in a Joukasou system at some degree, and a drain field to assimilate effluent water. Some septic systems are equipped with a further treatment process.

STORMWATER BEST MANAGEMENT PRACTICE: Control measures taken to mitigate the impact of stormwater in both quantity and quality. It could be structures constructed on site, or non-structures such as public education, land use planning or elicit discharge detection and prevention measures.

STRUVITE: Crystal of phosphate, magnesium, and ammonium ($NH_4MgPO_4 \cdot 6H_2O$), which can be used as slow-releasing fertilizer. Uncontrolled formation of struvite inside the wastewater treatment plants forms scales and clogs the piping system, impairing its performance.

CHAPTER 7

BIOGEOCHEMISTRY: Biological and geological influences on chemical forms and processes in ecosystems.

CRITICAL VALUE: The concentration of soil-available P, estimated by an extract (e.g., Olsen P), above which no significant increase in crop yield is achieved.

CULTURAL STIGMA: Something that is perceived by a group of people as something that is shameful or undesirable.

FECOPHILE: Cultures where human excrement is used to fertilize agricultural crops.

FECOPHOBE: Cultures where human excrement is not used to fertilize agricultural crops.

FIXED-P: Phosphorus that is tightly bound in an inorganic matrix that makes it unavailable for uptake by plants.

NON-FIXED P: Phosphorus that is readily available and can be taken up by plant roots.

TRANSGENIC CROPS: An agricultural plant where the genetic composition has been altered by directly inserting genes from another organism into the plant. A transgenic crop differs from a traditional crop in that in the latter, the genes have been altered through selective breeding rather than direct insertion of DNA.

CHAPTER 8

COMPLEX ADAPTIVE SYSTEMS: A complex adaptive system is a complex system viewed according to how it can change over time and space and in response to contextual factors.

COMPLEX SYSTEM: Characterizing a system as a complex system shows how causality goes beyond simple one-way connections. Feedback, time delays, and nonlinearities characterize relationships between some components and processes of a complex system.

EDUCATION REFORM: Education reform is a political process aimed at improving public education. Small improvements in education are thought to have large social returns in health, wealth, and well-being.

EPISTEMIC: Epistemic is an adjective that refers to the knowledge and belief development process. For example, an "epistemic barrier" is a knowledge-developing barrier, or something that prevents an individual or group from understanding something.

MATERIAL FLOW ANALYSIS: A Material Flow Analysis is a systematic assessment of the stocks and flows of materials within a system defined in space and time; it connects sources, pathways, and intermediate and final sinks of a given material.

SYSTEMS THINKING: Systems thinking involves understanding how components and processes influence one another within a whole. In nature, systems thinking involves looking at an ecosystem and how various elements such as air, water, plants, and animals, as well as trends such as climate shifts and population changes, determine whether populations survive or perish.

TRANSDISCIPLINARY (RESEARCH AND PROCESS): Research approach that links scientific theories and results with real-world data from practitioners. Transdisciplinary processes involve authentic collaboration among science and society and include representatives from industry, government, administration, different stakeholder groups, and the public at large.

CHAPTER 9

BACKCASTING: Determining the trajectory of change of a system from an envisioned sustainable future back to the current system state; determining the "path forward" to a sustainable future by starting at that future system state.

INTERVENTIONS: Purposeful action by an agent to create change in a system as it moves through a backcasted trajectory toward a sustainable future state; solutions associated with intervention points must be integrated.

STAKEHOLDER: Any person or group who can affect or is affected by a particular situation or system.

SUSTAINABLE SCENARIO DEVELOPMENT: Development of images or visions of possible future system states that are unconstrained by the inertia of the current system state.

TRADITIONAL FORECASTING: Projecting future trajectories of a system from the present or the past to determine a probable or expected future.

VISIONING: Crafting of future desirable states of a system that are unconstrained by the current state of the system.

INDEX

Acidic soils, 73, 93
Activated sludge process, 104
Africa, 10, 11, 44, 54, 56, 91, 94, 115, 117, 147, 153, 158, 161, 206
Agriculture, 12, 13, 23, 28, 29, 33, 34, 42, 43, 51, 52, 77, 89, 90, 91, 93, 97, 101, 102, 103, 105, 115, 117, 126, 135, 145, 149, 159, 160, 161, 167, 188, 191, 192, 193, 200, 201, 204
Algeria, 47
Alkaline soils, 73, 90, 149
Anaerobic digestion, 121, 129, 193
Apatite, 42, 43, 44, 46, 105
Argentina, 103, 203
Asia, 94, 147, 158, 206
Assimilation, 48, 66, 145, 150, 161
ATP, structure of, 26
Australia, 48, 121, 129, 190, 205
Auxin, 78

Backcasting, 185, 186, 187, 188, 189, 190, 209
Band application, 76
Best management practices (BMP), 73, 125
Bioavailability, 43, 75, 90, 192
Biofuel and biofuel production, 66, 68, 70, 71, 72, 73
Biogeochemistry, 145, 153, 157

Biosolids, 103, 104, 121
Brazil, 44, 47, 97, 150
Broadcast application, 76
Buddhism, 159
Burkina Faso, 147

Calcareous soils, 102
Canada, 48, 80, 94
Carbon
  cycle, 28, 153, 207
  element, 28, 30, 77, 78
  sequestration of, 134
  trophic transfer of, 153
China, 12, 48, 54, 56, 91, 117, 127, 147, 154, 155, 194
Closed (loop or system), 128
Communicative planning, 134
Complex adaptive systems, 171, 174, 176, 178, 179
Complex system, 171, 178, 179
Conservation Reserve Program (CRP), 70, 71
Constructed wetlands, 34, 125
Cost structure, 45, 58, 60
Critical value, 149, 150
Crop-based agriculture, 93, 97, 105
Cultural stigma, 157

Denmark, 135
Diagenesis, 44
Diet
   animal, 28, 79, 80, 81
   human, 10, 153, 154, 155, 156, 157, 161, 208
DNA, 24, 28
Dry distiller's grain (DDG), 71

Ecological sanitation (EcoSan), 127, 161
Education reform, 171, 178, 179
Effluent polishing. *See* wastewater and wastewater treatment process
Egypt, 50, 147
Element, 23, 28, 41, 43, 143, 152, 153
Energy
   required for human phosphorus cycle, 16, 45, 48, 57, 60, 114, 133, 145, 160, 193, 194
   required for wastewater treatment, 118, 119, 121, 126, 127, 129, 131,
Enviro-pig^{tm}, 79, 80
Epistemic, 178
Erosion, 17, 29, 33, 55, 60, 88, 89, 92, 93, 97, 113, 130, 131, 150, 152, 167, 192
Eutrophication, 10, 11, 16, 29, 33, 35, 54, 76, 79, 88, 91, 92, 93, 97, 100, 102, 103, 104, 106, 118, 132, 133, 149, 150, 151–152, 202, 203, 208
Excreta, 12, 67, 92, 101, 102, 117, 118, 119, 127, 136, 146, 157, 158, 159, 160, 161, 188, 192
Externalities, 5, 194, 195

FAO, 68, 90, 158
Fecophile, 157, 158, 159
Fecophobe, 157
Fertilizer, 3, 9, 10, 11, 13, 14, 20, 29, 30, 32, 33, 34, 42, 43, 49, 51, 52, 53, 54, 55, 57, 59, 60, 65, 66, 67, 68, 70, 72, 73, 75, 76, 77, 88, 90, 91, 92, 93, 94, 97, 98, 100, 101, 102, 103, 105, 114, 117, 121, 124, 126, 130, 134, 143, 149, 150, 151–152, 157, 158, 160, 168, 173, 176, 177, 178, 179, 180, 184, 185, 188, 189, 191, 192, 193, 194, 200, 201, 204, 205, 206, 207, 209
Fertilizer phosphorus, 73
Fertilizer phosphorus-use efficiency (FPUE), 66, 73, 75
Finland, 44, 135
Fixed phosphorus, 150
Food production, 2, 12, 14, 15, 16, 66, 88, 102, 105, 106, 114, 117, 129, 130, 136, 143, 146, 160, 188, 191

Food security, 9, 14, 15, 16, 145, 147, 160, 170, 188, 204, 206
Food supply chain, 114
Footprint, *as in* "P footprint," 193, 208, 209
Forecasting, 57, 59, 185, 186, 187
France, 54, 121, 147, 174

Genetic engineering, 2, 30, 79, 146, 147, 150, 152, 153, 161
Genetically engineered plants, 162
Genetically modified crops, 147
Genetically modified organisms, 162
Germany, 131, 135, 147
Governance, 8, 16, 170, 187, 207, 208
Green building, 127, 133, 135, 160
Green Revolution, 13, 14, 54, 94, 206

H+PPase, 78
Health (human), 146, 147, 187
Holistic perspectives, 185
Hubbert curve, 58
Human phosphorus cycle, 188
Human wastes (or excreta), 29, 105, 117, 118
Hydroapatite, 124
Hydroxyapatite, 26
Hypoxic zone, 35

Igneous phosphate ores, 44
Immobilization, 32
Improved sanitation, 118, 160
Industrial symbiosis, 117
Intermediate product, 53
Interventions, 15, 16, 189, 190, 205, 209
Israel, 48

Jainism, 154
Japan, 117, 135, 147, 158, 173
Japanese Society, 159
Jordan, 48

K+ (potassium), 75
Kazakhstan, 48
Kenya, 94, 147

Land use, 52, 68, 89, 91, 93, 102, 105, 153
Lime, 103, 119, 121, 127

Manure and manure phosphorus, 12, 13, 28, 33, 66, 67, 70, 71, 73, 75, 76, 79, 81, 88, 91, 93,

94, 97, 98, 99, 100, 101, 105, 117, 118, 125, 131, 130, 132, 134, 157, 188, 192, 193, 195
Markets, 45, 60, 94, 105, 193, 207
Material flow analysis, 171, 172, 179
Mineralization, 28, 32
Mining, 2, 3, 10, 13, 14, 29, 42, 43, 44, 45, 46, 47, 48, 49, 53, 54, 55, 57, 58, 60, 143, 185, 188, 191, 193, 194, 206
Monogastric, 98
Morocco, 49, 50, 54, 117, 194

Namibia, 48, 194
Netherlands, 35, 115, 117, 129, 132
Nitrogen
  cycle, 28
  element, 20, 21, 29, 76, 79, 121, 153, 193
Non-fixed phosphorus, 150
Non-ruminant animal, 26, 28
Norway, 102, 103, 135
Nucleic acids, 23–24
Nucleotide, structure of, 26

Occlusion, 12, 30
OCP, 49
Omnivore, 153, 157
Open (cycle or system), 12
Ore, 43, 45, 47, 49, 52, 53, 60, 114, 124, 173

Pedogenesis, 12
Phoenix, Arizona, 102, 190
Phosphate, structure of, 23, 24
Phosphate rock, 43, 44, 45, 46, 47, 48, 49, 50, 51, 53, 54, 56, 57, 58, 59, 72, 75, 88, 93, 192, 194, 204, 205
Phosphatidylcholine, strucutre of, 25
Phosphorite, 23, 43, 44, 45, 47
Phosphorus
  accessibility, 11
  assimilation efficiency, 28, 145, 150, 161
  balances, 94
  element, 23, 41, 152
  management, 177
  recovery, 112, 207
  supply chain, 135, 177, 205
  transfer continuum, 33
Phosphorus use efficiency (PUE), 66, 67, 68, 76, 77, 78, 79
Photosynthate, 77
Phytase, 28, 79, 80, 81, 98,

Phytate, 26, 28, 79, 80, 81, 98
  structure of, 25
Polymictic, 104
Polyphosphate, 26, 75, 121
  structure of, 25
Population (human), 6, 10, 12, 13, 14, 55, 56, 58, 59, 101, 102, 170, 188, 190, 191, 203, 204, 206
Population equivalent, 102
Posphate rock decision support system (PRDSS), 75
Problem-solving framework, 5

Quality of water, 10, 32, 34, 35, 71, 102, 104, 105, 106, 125, 130, 203, 205

R/C ratio, 49, 58
Recoverable concentrate, 49, 51
Religion, 152, 154, 155
Renewable energy, 68, 131, 133, 136
Republic of South Africa, 44, 147
Reserve or ore reserve, 45, 46
Reserve base or resource base, 32, 45, 46, 47, 48, 49, 51, 53
Reserves of mineral phosphorus, 43, 45, 46, 47, 48, 49, 50, 51, 52, 53, 58, 59
Resource pyramid, 45, 47, 48, 49
Rhizosphere, 26, 32, 78, 150, 152
Riparian buffers, 106
RNA, 24, 28
  structure of, 24
Roosevelt, President Franklin D., 205
Ruminant animals, 71, 155
Russia, 44, 48, 54

Sanitation, 12, 118, 126, 127, 133, 159, 160, 161
Septic system, 124, 125, 160
Sewage sludge, 103, 119, 121, 125, 130
Social-ecological system, 6
Soil phosphorus, 77
Sorption capacity of P, 90, 94, 97, 105
Stakeholder, 5, 15, 105, 135, 172, 174, 176, 177, 179, 185, 188, 190, 194, 209,
Stormwater, 125, 133, 135
Stormwater best management practice, 125
Stover, 67, 193
Struvite, 121, 126, 133, 134, 160, 193, 206
Subsoiling, 101
Sustainability, definition of, 5

Sustainability science, definition and framework of, 3
Sustainable Phosphorus Summit, 205
Sustainable scenario development, 185
Sweden, 106, 124, 126, 135, 205
Syria, 48
Systems thinking, 3, 16, 172, 176, 178, 179

Trade, 14, 188, 207
Traditional forecasting, 185
Transdisciplinarity, 176
Transdisciplinary, research and process, 15, 170, 171, 176, 177, 178, 180, 189, 209
Transgenic crops, 146, 151, 152
Tunisia, 48

United Nations, 5, 51, 207
United States of America, 10, 12, 35, 49, 52, 54, 56, 59, 68, 71, 72, 93, 94, 99, 102, 104, 119, 124, 125, 141, 147, 150, 155, 156, 157, 162, 194

Urban systems, 92, 93, 179
USDA, 157
USGS Mineral Commodity Summary, 49, 51, 52, 54, 58

Variable rate fertilizer application (VRA), 77
Vegetarian, 10, 101, 153, 154, 157, 192
Visioning, 185, 188, 190

Wastewater and wastewater treatment process (advanced and conventional practices), 26, 35, 92, 93, 101, 102, 106, 118, 119, 121, 124, 133, 135, 202, 203, 206, 209
Water, as a limiting resource, 145
Western Sahara, 50
Wicked problem, 6, 9, 17, 202, 205
World Bank, 206, 207

X-Rays, 27

Zimbabwe, 44, 147